ORNAMENTAL WATERFOWL

A Guide to Keeping and Breeding

ORNAMENTAL WATERFOWL

A Guide to Keeping and Breeding

Richard Brigham

BLANDFORD

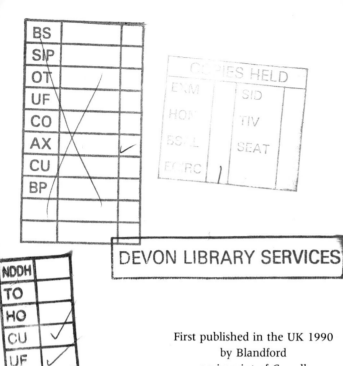
First published in the UK 1990
by Blandford
an imprint of Cassell
Villiers House, 41/47 Strand, London WC2N 5JE

British Library Cataloguing in Publication Data
Brigham, Richard
Ornamental waterfowl.
1. Ornamental Wildfowl. Care & Breeding
I. Title
636.68

ISBN 0–7137–2187–1

Typeset by Litho Link Limited, Welshpool, Powys, Wales
Printed and bound in Great Britain by
Biddles Ltd., Guildford and King's Lynn

Contents

To my parents,
whose home became a menagerie
in my early birdkeeping years.

Preface

I HAVE BEEN OBSESSED with a passion for bird-keeping for as long as I can remember. It began with a succession of orphans in the shape of various owls, magpies, jackdaws and rooks, which every country boy at the time seemed to accumulate in his back yard. Gradually it was supplemented by a seemingly inexhaustible supply of injured waifs and strays as diverse as house martins and herons, that turned up on my doorstep, brought in by well-meaning friends and for the most part afflicted with broken legs and wings that I 'treated' as best I could. In spite of, or possibly as a direct result of, my treatment almost everything hopped or limped about the place, as my rather limited veterinary skills of that time left much to be desired. Anything with the full use of both its legs and wings and other faculties was certainly the exception rather than the rule.

My introduction to waterfowl occurred one morning in the shape of 'Hoppy', a young Canada goose with a noticeable limp discovered wandering aimlessly about the marshes, and he lived with us for many years during my schooldays. Each evening Hoppy would hear the clanking of my bicycle along the road as I returned from school, and would honk loudly long before I reached the garden until I went to visit him in the large aviary where he spent the day confined, and which he shared with a pair of kestrels. He would follow me around the garden when I let him out to graze on an area of grass in front of the stables. Being a youngster, Hoppy become hopelessly imprinted on me, honking a greeting whenever I appeared in sight and talking to me constantly while I fed the kestrels in the aviary. And so began what was to become a lifelong interest in waterfowl, though it was long after Hoppy's demise that circumstances prompted me to take a serious interest.

It all began one cold winter's morning in late January as a close friend and I were pottering around the marshes close to my home. I am lucky enough to live on a farm, with the freedom to roam and take delight in the many aspects of wild life which its varying habitat provides, particularly the low-lying marsh-lands bordered to the north by the grey arm of the river and containing a huge lake of many acres which attracts all manner of fowl. Wandering beside an area of rough, boggy ground alongside the river, a wasteland of almost impenetrable vegetation and a jungle of dead and decaying trees, we disturbed a flight of mallard from the more open of a chain of three silted-up ponds enclosed by the trees. Although normally dried out and nothing more than a morass of mud and debris, at the time they held a few inches of floodwater drained off from the surrounding fields following several days of torrential downpour. The place had been neglected for years, a forgotten corner and of no use as farmland. It seemed such a waste. Making our way along the bank separating the ponds from the river, I realized that the mallard had sparked off an idea and at last I could visualize a use for the place. It seemed an ideal situation for setting up a waterfowl collection.

But there was obviously a great deal of work to be done if the idea was to reach fruition. The entire area – an acre or so in extent – had previously been used as a chain of stew-ponds for rearing trout. Having lain dormant for almost 60 years it was now nothing more than a completely overgrown jungle of neglect, with a few inches of water – albeit temporarily – concealing the mud. With a little imagination the place looked an ideal site for development. Even now, great beds of interwoven rushes encroached completely across two of the ponds, thick, quaking rafts supported above a glutinous, stagnant, pudding-like mess filled with decades of fallen branches and rotting trunks of elm, alder and blackthorn scrub that had accumulated over the years. Despite this, if the place could be cleared and cleaned and returned to something approaching its former glory, it had an obviously unlimited potential.

But where to begin? It seemed a daunting task. A forest of massive, dead elms leaned drunkenly across the water, victims of the deadly Dutch elm disease that was rife at the time. Years of impenetrable blackthorn scrub had sprouted up to choke the banks, interwoven with tangles of elder, massive bramble patches, beds of nettles and decaying alders; the whole scene a picture of waste and neglect. And just how were we to set about clearing it? The steeply sloping banks running down to the water would prohibit the use of anything mechanical over much of the area, so it would be largely a matter of laborious spadework.

Luckily Ken, a lifelong friend with a shared interest in the outdoor life, had never been afraid of hard graft, and also welcomed a bit of a challenge. Hardly an hour had elapsed before we were back with a shovel apiece, ready to set to work and at least see what could be done to the actual ponds.

And a challenge it was too! After several hours of back-breaking toil we were both sweating profusely despite the cutting wind. We were liberally spattered with mud from the exploratory dig, having shifted several tons of black, oozing mud without seeming to have made a great deal of progress; though by the end of an exhausting day we had succeeded in cutting a new outlet channel from the lower downstream pond into the river, and the floodwaters were at last beginning to move.

Returning early next morning the pond was almost dry. We began to work gradually upstream, cutting and gouging a rough channel with a view to eventually reaching the top inlet drain and getting a flow of water moving. Progress was sickeningly slow, though we set to with obsessive enthusiasm, ending the second day thoroughly exhausted but satisfied with our labours.

This set the pattern for the next few weekends, when every moment of our spare time was occupied in the backbreaking task of digging, clearing, cutting and shovelling. Eventually, we had dredged out a central channel throughout the entire length of the three ponds, a distance of well over 100 yd (100 m), about 3 ft (90 cm) wide by 2 ft (60 cm) deep, already encouragingly beginning to fill with water. The ultimate success of the venture depended on the water levels; whether or not there was a sufficient fall throughout the length of the ponds to achieve a healthy flow of water and bring up the level of the ponds when the upstream inlet was re-opened and the downstream outlet dammed.

The day of testing came when we constructed a rough dam to alter the course of the water upstream, pile-driving wooden stakes into the stony bed of a stream near the top inlet channel, and fixing rough planks across the water to cut off the stream's normal route to the river, hoping instead to restrain the flow sufficiently to raise it the 6 in (15 cm) necessary to divert it through the ponds on its original course, which had been dried out for so many years. This was the moment of truth. Would the water flow, or were our efforts all in vain? It was a long and anxious wait, the water taking several hours to build up to a level that would force it through the ponds. We were delighted when it at last began to move, a mere trickle at first that built steadily into a slow but constant flow before the level of the drain reached high enough to breach our makeshift dam. This was all we wanted to know. Our plan, it seems, was beginning to work.

We set to with renewed enthusiasm. The flow itself was also beginning to work wonders, slowly but surely moving the stirred-up deposits of silt as it passed through to the river. the process was hastened a couple of weeks later when we precariously pumped out a few loads of silt with a tractor and slurry cart, and gradually over the weeks the pool of water began to widen to its original contours. Each pond, about 30 × 10 yd (27 × 9 m) was separated from the next by a narrow channel edged with parallel brick walls. During its use as a trout-stew the walls were rebated to allow a system of planks to be inserted to control the water at an ideal level, and to build it up to form a waterfall at the inlet end which could be used to oxygenate the water periodically for the trout. The planks had long since rotted, but these I replaced with welded mesh gates made up by the local blacksmith, the 2 in (5 cm) mesh being sufficient to allow the water to flow but small enough to prevent any fowl escaping.

Having obtained a reasonable amount of water, our next job was to erect a fence. The ideal would be a minimum of 6 ft (1.8 m) with an outside overhang to discourage access by predators, but discovering the prohibitive price of new wire netting, we decided that such a luxury was for the moment out of the question. Having just set myself up in business as a taxidermist, my available money was already tied up and being temporarily on the breadline I was forced to settle for a load of discarded wire netting from a line of demolished aviaries at the local wildlife park where I worked a few hours in my spare time. The netting had been rolled up in a tangled heap and stacked ready for consignment to the rubbish dump. It was in a sorry state, but we salvaged what we could, and after spending an entire morning straightening and repairing the best of it, we had sufficient to encompass the first of the three ponds, the smallest, making it reasonably secure, though the completed fence seemed to pose little deterrent to a determined fox seeking a closer relationship with any inmates. On such a limited budget it was the best I could hope for.

The most important part of the venture was the acquisition of the waterfowl themselves, and here I was also forced to start out on the bottom rung of the ladder, acquiring a small clutch of eggs from a mallard nesting beside the farm lake. That the duck had obviously been on intimate terms with a white domestic drake become evident only when the half a dozen eggs, placed under an obliging broody hen from Ken's laying fowl, duly hatched. It was a motley crew of variously coloured ducklings that at last followed the clucking hen from the confines of the hatching coop. But they would have to suffice, at least giving me the chance of learning by experience the basic rudiments of duck husbandry before progressing to more difficult and therefore more expensive birds.

Typical of mallard, the six ducklings all thrived and duly grew to adults, and it was an exciting day indeed when we transported them from the garden pen in a ricketty crate and released them on the pond, a transition they obviously enjoyed immensely, splashing and diving with gay abandon until they had explored every inch of their new surroundings, and eventually settling to preen on a log overhanging the water. It was a joy to watch them. And so began an all-absorbing interest that was to last for years.

But the hard work was by no means finished. For many months, each weekend – come rain or shine – was spent working at the ponds, cutting, felling, logging and clearing the jungle of dead elms, alders and scrub with the aid of a chainsaw, dredging more of the ponds and ending each day absolutely exhausted, cut and scratched by brambles, eaten alive by flies, reeking of smoke from countless bonfires and liberally plastered with mud from head to toe. Almost always one or other of us managed to fall in, particularly when attempting to remove boughs and branches from the water up the steep slippery banks. The felled tree-trunks proved the biggest problem. We removed what we could with a pair of grappling hooks on a rope, but had to enlist the use of a tractor and chain to haul the massive trunks clear of the water, where they could be logged into manageable lengths for firewood. It was hard though enjoyable toil, but progress was good and it was not long before my thoughts turned to acquiring more fowl to increase the collection.

Sadly, my next attempt at rearing proved a disaster. Having been lulled into a false sense of security by the ease with which the mallard brood was reared, I was ill-prepared for the difficulties that arose with my first batch of tufted ducks. Obtaining a clutch of eggs later that summer, I was pleased when the majority hatched on time, but they proved extremely difficult to rear and, unlike the more robust mallard, declined to feed almost immediately upon hatching. The tiny, sooty black ducklings succumbed one by one over the next few days until there was but one solitary survivor, a tiny female who, although appearing to be the runt of the clutch, somehow managed to outlive her siblings and was eventually released safely on the pond. In fact she lived for many years, a lively little character that dived incessantly in the deeper regions of the pond and grew quite tame over the years. The experience taught me a lesson. There was obviously much more to the rearing game than I had thought.

By the following autumn I had become more ambitious, laying out an amount of hard-earned cash to obtain a pair each of pintail, shelduck, Chiloe wigeon and carolinas from a breeder, all of which survived the winter to breed the following spring, by which time I had learned a little more of the rearing process. Selling the surplus enabled me to invest in yet more new species.

And so it has continued to the present day. The pond complex and its inhabitants have progressed beyond all expectations, to a total of well over a hundred stock birds of more than forty species, many of which breed freely to provide a surplus for selling or swapping, to enable me to acquire even more new and exotic types of fowl, and helping with the running costs of the venture. Thus it provides an almost cost-free and delightful hobby that absorbs much of my spare time and provides a constantly changing pattern of interest throughout all seasons of the year. Widened out to their original contours, the ponds are now supplied with a constant flow of fresh water that attracts all manner of bird, amphibious and insect life; ample reward in itself when compared to the sterile and decaying mess of a few years ago. Even shoals of fry and stickleback venture upstream from the river, their quicksilver forms twinkling in the shallows where they are pursued by my little flock of hooded mergansers, which thrive on such a tasty addition to their diet, and also tempting in the local kingfishers to feed. On warm summer days the place is alive with vivid dragonflies, snapping up flies over the water like miniature predatory helicopters; and butterflies now visit in huge numbers, pausing to drink at the stand of purple thistles left for that very purpose. Crayfish return sporadically, but are taken by the eider ducks, who consider them a real delicacy, crunching them in their powerful beaks after diving to capture them on the muddy bottom. Clumps of wild flowers now have the chance to bloom and set their seeds; primroses, bluebells, red and white campions and many more, some introduced after the clearing was completed but others, inevitably, turning up miraculously now that the conditions are to their liking. Over the years I have compiled a rough list of visiting birdlife attracted by the water and its amenities, a list that contains such welcome visitors as oystercatchers, green sandpipers, redshanks, water rails and no less than eleven species of wild duck, some obviously escapees from other collections. All in all the hours of back-breaking work have been very worthwhile, and the project has been a most rewarding experience.

For this reason I hope the following chapters may in some small way inspire a few of you who are lucky enough to have a disused pond or stream on your land, or the facilities for creating one, to keep a few head of waterfowl, whether it be a backyard effort with just a couple of pairs of fowl or a more ambitious venture on a large pond or lake. And I hope to guide you along a straighter path than that which I took, helping you to avoid at least some of the obstacles and pitfalls I encountered on the way. I cannot claim to be anything of an expert, simply someone who enjoys keeping and rearing fowl, and who has acquired a little knowledge on the way. What I have learned is that almost anyone with a garden can keep a few waterfowl, and there is much to recommend it. In a suitable environment, birds will soon become quite tame, though they are always active and on the move, whether it be feeding, diving, preening, bathing, catching flies across the water or indulging in the untiring delights of their fascinating springtime courtship displays. Waterfowl are an ideal subject for the bird-keeper and one of the most fascinating groups in the bird world, for years having stirred the heart of hunter and conservationist alike. In captivity they require the minimum of attention, will keep themselves in peak condition on good water, and many will earn their keep by producing a yearly batch of saleable offspring. Thus it can be seen that a little initial effort and expenditure will be, in many ways, amply rewarded.

1

Starting a Collection

With some breeders offering a baffling variety of well over 100 species of fowl from the 240-odd species spread throughout the world, it will doubtless seem confusing to the complete beginner, who needs to know where to start, and which species can happily be kept together in the best environment he is able to provide. Any reputable breeder will be only too willing to advise, but as a rough guide one should always bear in mind that the most important requirement for all web-footed fowl is water, its quality, quantity and depth dictating to a great extent the types and numbers of birds one is able to keep, whether it be on a natural water or an artificial pond.

Apart from the geese and swans, the majority of waterfowl can be divided into two main categories, the surface-feeders or 'dabbling' ducks, which are quite content if provided with just a few inches of water, and the 'diving' ducks, which in their natural wild state obtain most of their food by diving. On natural water of a reasonable depth almost anything can be kept, particularly if the pond is fed by a flowing stream, for such water is a breeding ground for myriads of aquatic creatures and as such harbours plenty of natural food. On an

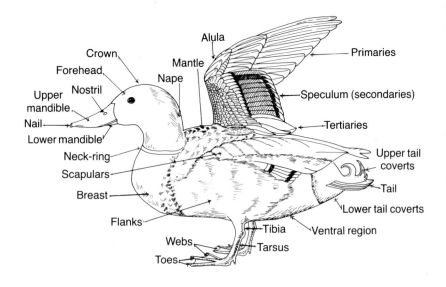

Fig. 1 Topography of a duck

artificial pond of limited size one should perhaps restrict oneself to keeping mainly dabbling ducks, as the divers are seldom content unless provided with at least 2 ft (60 cm) of water and will do much better on a diet that is supplemented by the minute water creatures, insects, larvae and plant-life that are always to be found in healthy natural water.

If one has the good fortune to own a natural pond, lake or widened-out stream, so much the better, for very little will need to be done apart from erecting a secure boundary fence, landscaping the interior and possible planting a few low shrubs and conifers to act as shelter belts in adverse weather.

It is also important to ensure that there are sufficient shallow banks to enable fowl to get on and off the water easily to feed. This is particularly vital in the case of diving ducks, which walk with some difficulty on dry land. The banks and water margins can be planted with various indigenous rushes, sedges and reeds, though these will need protecting from the birds with small-mesh wire-netting surrounds until they become well established. Besides creating a natural-looking habitat, in time the plant-life will encourage a whole host of water creatures to move in to add to the natural food supply, as well as providing added interest. In due course many varieties of marsh plants will also produce titbits for the fowl in the form of ripened seedheads. In short, the ideal natural pond should be of varying depth, be reasonably sheltered from the cold northerly winter winds and be well planted with vegetation. An island or two on a larger water is a distinct advantage, providing safe and secluded resting spots and ideal nesting sites if sufficient cover is provided.

The majority of would-be waterfowl-keepers are less fortunate, and will need to create a pond from scratch, constructing it from either concrete, a PVC pool-liner, or possibly a pre-cast fibreglass or plastic pool, which is quite adequate if one is keeping only a few pairs of fowl. For their size pre-cast pools are quite expensive, but one has the advantage of an instant pond that simply requires digging into the ground and filling with water. PVC pool-liners are also far from cheap, though they will allow one to be a little more imaginative in the construction of a pond as their design is not unduly restricted.

Almost certainly the cheapest and most satisfactory artificial pool is one constructed from concrete, which can be built to any size, shape or design. A pool of any size needs to be at least 4–5 in (10–12 cm) thick and its construction initially requires a bit of hard graft both in preparing the ground after digging out the pond, and in mixing and laying the concrete. After excavating the pond and ramming a good layer of hardcore firmly into the base, the concrete is best laid on a bed of wire netting, which both reinforces the mixture and helps hold it in position for drying where the surfaces are sloped. When worked to a smooth finish and left to dry completely the surface can be coated with a pond-sealer. Whatever its depth, at least one side of the pond should be gently shelved off and ribbed to allow the fowl to climb out easily, and finished off with a wide apron of concrete at ground level to separate the water's edge from the surrounding ground, as ducks love dibbling around in mud and once the ground becomes wet will soon churn it into a veritable quagmire, making the water muddy as they paddle constantly in and out of the pool. Instead of featureless concrete, the apron can be made much more attractive by using crazy paving or natural stone.

By whatever method and with whatever material the pond is built, it is important to remember that the water will need changing regularly as it becomes soiled, and some method of emptying and cleaning the pond easily and quickly is required. In the case of a pre-cast pool the water will probably need to be baled out, or siphoned off with a hose-pipe, and thoroughly cleaned when almost empty, whereas with a concrete pond a drainage system can be incorporated when building, allowing quick and easy emptying by means of an underground soakaway. When building my own garden pond where most of the young fowl are hardened off, I included a system of two underground tanks connected to the pond outlet by a drainpipe (Fig. 2). When the bung is removed the water passes first to a large

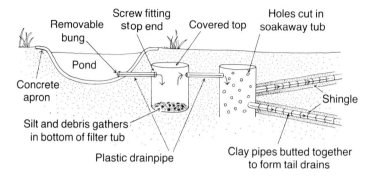

Fig. 2 Soakaway system

covered plastic water butt sunk in the ground, which when full has an outlet at the top to drain off and carry excess water to the actual soakaway. Most of the mud, dead leaves and other debris that has accumulated in the pond is left in the first tub, allowing the water to percolate into the ground from the soakaway without clogging, as would be the case if everything was allowed straight into the soakaway. The soakaway itself is simply another water butt turned upside down in the ground, in which holes have been cut at different heights to allow the water to escape. To enable the water to drain away faster, a couple of tail drains can be run from the soakaway to distribute the water over as large an area as possible. The tail drains are best made of short lengths of clay pipe laid at the bottom of a trench on a bed of shingle, the pipes simply butted against one another to allow water to escape between the joins along the complete length of the drain. These drains should be made as long as possible to facilitate faster emptying of the pond, the pipes covered with a few more inches of shingle and finally with polythene sheeting to prevent soil from washing down to clog the pipes. The first filter tub has to be emptied occasionally as the silt builds up, but this is no problem. The system works well, allowing the pond to be emptied and cleaned quite quickly, sweeping all the debris into the outlet pipe with a broom to be separated in the filter tank.

Of equal importance for a good supply of water is a secure perimeter fence. The siting of the enclosure will dictate what type of fencing is necessary, for it is not so much a matter of simply keeping in the fowl – in most cases 2 ft (60 cm) would suffice to restrict pinioned birds – but rather more importantly of keeping out vermin attracted by the contents of the feeding bowl, and the fowl themselves, which provide temptation for predators such as foxes, stray cats and dogs, stoats, weasels and other equally undesirable visitors. When fencing it is a matter of weighing up the dangers of your particular locality and building accordingly.

In open country the ideal fence would be a minimum of 6 ft (1.8 m) high, with two overlaps of wire on the outside to discourage vermin, one at the top to prevent the more adventurous from climbing over and, if two different sizes of wire mesh are used, another a couple of feet from the ground to prohibit small vermin from gaining access. A little attention to detail when building the fence will pay dividends later, for once all traces of ground vermin have been eradicated inside the pen there should be few problems in the future.

Good strong posts of oak, ash or elm are to be recommended as they last much longer than the cheaper softwood conifers, which will quickly rot at ground level and eventually snap off. The posts should be stripped of bark and dipped in preservative before being erected, and given a regular coat of timber preservative at least once a year. Metal angle-iron posts are a good alternative, and although initially more expensive will last for many years.

The most expensive outlay will be the wire netting. Galvanized wire netting or chainlink fencing is far from cheap, and the smaller the mesh, the more expensive it becomes. For this reason it is better to use two sizes, a few feet of ½ in (1.25 cm) mesh at the bottom to prevent entry of rats and other small pests at ground level, filling in with larger mesh at the top, which cuts down the overall cost considerably. The lower run of netting should be about 3 ft (90 cm) high and buried at least 1 ft (30 cm) underground, curving outwards to prevent anything from tunnelling under the wire. The top of this lower section can be bent outwards and downwards a full 6 in (15 cm) to prevent rats from scaling the small mesh and gaining entry higher up where the mesh widens. The top portion of the fence can be completed with 2 or 3 in (5 or 7.5 cm) mesh, the two runs overlapped and fastened together with netting clips. The very top of the fence needs a good overhang of at least 1 ft (30 cm) but preferably 18 in (45 cm), over which nothing will be able to climb. Such a fence (Fig. 3) will stop almost anything on four legs, but without the overhang a plain vertical fence is no deterrent to a hungry fox, be it 6 ft (1.8 m) or even higher, for foxes possess the agility of a cat and will scale wire netting with ease, as I once found out to my cost. I lost a female barnacle goose one night as the pair roosted beside the water, and by following a tell-tale trail of feathers I found where the fox had gained entry over the vertical 6 ft (1.8 m) fence. What was even more remarkable was that it climbed out again carrying the dead goose in its mouth. This it obviously did, for when tracking it later I discovered the almost intact remains of the goose, minus its

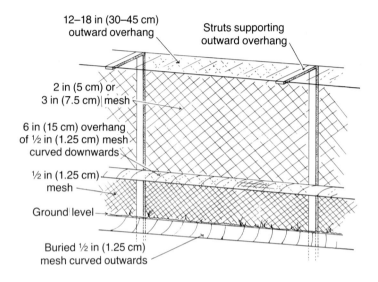

12–18 in (30–45 cm)
outward overhang

Struts supporting
outward overhang

2 in (5 cm) or
3 in (7.5 cm) mesh

6 in (15 cm) overhang
of ½ in (1.25 cm) mesh
curved downwards

½ in (1.25 cm)
mesh

Ground level

Buried ½ in (1.25 cm)
mesh curved outwards

Fig. 3   Perimeter fence

head, carelessly half-buried on the nearby marsh where it had been stored for a future meal. The fox tried the same trick a few nights later, but I was not caught out again.

If, however, one lives in an environment where foxes pose no problems a lower fence will suffice, but it should still be at least 4 ft (1.2 m) high with a generous overhang to prevent marauding cats from jumping or climbing over, for a seemingly harmless moggy can leave a trail of devastation among a flock of tame, pinioned waterfowl.

All perimeter fences should be kept clean and free of vegetation, particularly on the outside, for rats and other small vermin dislike remaining for too long on open ground, where there is always the likelihood of being discovered by a hunting owl or hawk. Clean fences will discourage their

24

attentions and any well-worn tracks will be found by regular inspection.

Every enclosure will obviously need an entrance gate. This can often be the weak spot regarding vermin, but with a little foresight the problem can be easily overcome by building a raised swing gate (Fig. 4) hinged at the top and rat-proofed in the same way as the rest of the fence.

Any dividing fences within the main enclosure can be sectioned off with 3 ft (90 cm) netting of 2 in (5 cm) mesh, which is quite adequate to keep fowl from straying, and at this height a simple stile is easily erected to enable one to pass from one pen to another without needing to open a gate.

Apart from providing water and a secure boundary fence, very little else in the way of equipment is needed apart from various feeding buckets and bowls, a good mouse-proof food storage-bin and in due course a selection of nest-boxes and baskets. Waterfowl require little in the way of shelter, and will almost certainly refuse to use a hut even if supplied with one, though an open-fronted shelter can be useful for feeding the birds in snow or bad weather. Windbreaks of dense conifers and shrubs are sufficient to provide refuge from cold winds, and can be planted in such a manner that they both afford shelter and add much to the pleasing appearance of the enclosure. They will also be used in due course as cover for nesting, particularly if plants that provide a certain amount of ground cover are used. The remainder of the ground is best planted with grass, which should be trimmed short, except where rough areas are left in the spring for nesting, as many birds need and enjoy good short grass for grazing, especially geese and wigeon species.

Assuming one is feeding a mixture of pellets and wheat, the latter can simply be thrown in the shallows of the pond,

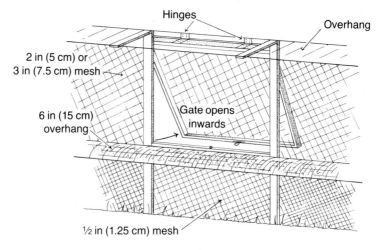

Fig. 4   Raised swing gate

which stops the sparrows from stealing it, but feeding bowls should be used to prevent pellets becoming fouled or trodden into the ground. Pellets thrown into the pond will soon disintegrate and foul the water.

Some of the more timid diving ducks, which are awkward and ungainly on land, are often reluctant to leave the water to feed. A diet of wheat alone is no good for them, and to ensure they get their fair share of pellets a few old frying-pans or shallow saucepans scrounged from the kitchen make ideal feeding bowls, the handles stuck firmly into the bank just above water level at a height which the birds can reach.

Last but by no means least, an essential part of the waterfowl-keeper's kit is an angler's landing net, an invaluable aid for catching up birds when moving them around or periodically checking their condition. One with a long handle and a deep, baggy net best fulfils the purpose, and can be purchased from any fishing-tackle shop quite cheaply. A net makes catching up an easy operation; far better than chasing birds around and destroying their confidence in you by repeatedly trying to catch them with your hands.

Having prepared a suitable environment and decided on which type of birds to keep, it is time to consult the breeder. A personal call is always far better then ordering by post or telephone, for one has the chance of giving the birds a quick once-over to check their condition, and most breeders will be ready to impart much helpful advice on their keeping. The best time to buy is in the autumn, as soon as the season's young stock have fully feathered, when most birds are sold as being hand-reared and pinioned – HRP for short.

When purchasing, always check that each bird has plenty of flesh on the breast, for loss of weight is often the only indication that an immaculately plumaged bird is ill. All birds should also have a bright, full eye, a healthy waterproof sheen to the feathers, and be cleanly pinioned. The tiny feathered alula should be left intact on the tip of the pinioned wing as it will cover and protect the point of pinioning which should at this stage be covered by feathers and virtually invisible. Incorrectly pinioned birds, particularly those with the alula removed by careless use of the scissors, will be liable to damage the severed joint when flapping around, and in the worst cases the wing-tip becomes bare and swollen, and subject to bleeding and infection if knocked.

Always endeavour to obtain unrelated pairs if possible, for although birds from the same clutch will usually breed just as readily, if the interbreeding is carried on through a few generations problems can develop, the most common of which is the subsequent production of infertile eggs. Any ducklings raised may display a variety of weaknesses, and are often difficult to rear. Regrettably there is no way of checking lineage, but if one deals with a reputable breeder who has a large stock of breeding birds and regularly buys in surplus youngsters from collectors at the end of each season, there is a better likelihood of his being able to supply truly unrelated pairs.

When introducing your first batch of waterfowl to an empty enclosure, always ensure there is plenty of food readily available at several points around the pen, the most important job at this stage being to get them all feeding well as soon as possible. If the enclosure is very large it may be safer to erect a small makeshift pen within its boundaries in which to house the birds for the first few days, otherwise some of the more timid ones may disappear amongst thick cover to skulk and refuse to come out for food. Penning for a day or two will allow them to become accustomed to their new surroundings, and you can check that they have settled down and are feeding well. They must, however, be released where there is access to plenty of bathing water after travelling, for after being cooped up on the journey – perhaps for several hours – they like nothing more than a good wash and brush up and the chance to bathe and preen bedraggled plumage back into order. Once a few fowl have become established, there should be no further problems when introducing new stock, as they will soon join forces with the residents and learn their way around.

Waterfowl, even relative youngsters, often carry quite heavy infestations of internal parasites in the form of worms, but these seldom cause problems unless a bird loses condition through some other factor, when a heavy infestation can add to the severity of the illness. Newly acquired stock birds can be treated with a worming powder, normally a flour-like substance which adheres well when mixed with pelleted food. Various brands already incorporated in the feed can also be purchased from some food suppliers.

2

The Control of Vermin

Wherever any form of livestock are kept in numbers they will attract an accumulation of what can only be classed as vermin, moving in to take advantage of a free and easy supply of food. Waterfowl are no exception and will attract a whole host of hangers-on, pilfering food supplies, causing destruction of nests and eggs and even preying on the fowl themselves if given the opportunity. Much of the problem can be eliminated by a secure boundary fence, but even then a few trouble-makers will inevitably turn up from time to time, particularly during the breeding season, when a squadron of winged predators are always on the lookout for a carelessly concealed nest or brood of young ducklings. Carrion crows, magpies, jays, jackdaws and rooks are the main culprits, but even the odd rogue moorhen or coot is not above robbing a nest or killing baby ducklings. The now common grey squirrel is also fond of eggs, and any living in the vicinity of a waterfowl collection should be dealt with harshly, and even old Prickles the hedgehog will not turn his nose up at such fare, though he can be taken alive and moved outside the pen, where he will cause no further trouble. Being a waterfowl-keeper, one must also become something of an amateur gamekeeper, learning to recognize the first signs of vermin moving in and the ways of keeping predation to a minimum.

Of the entire list of vermin harmful in one way or another to waterfowl, the brown rat is undoubtedly the worst. Labelled public enemy number one, it causes more wholesale damage than any other pest. It goes without saying that every opportunity must be taken to eradicate it by every available means: trapping, poisoning, shooting, snaring; and if this is not enough, by gassing the burrows if an infestation becomes severe. Doe rats are almost perpetually pregnant, a young doe being capable of breeding at a few weeks of age. It has been said that, given optimum breeding conditions, a single pair of rats can multiply to produce a staggering total of many thousands by the end of a single year. Being such prolific breeders, once a few rats are allowed to establish themselves they can prove the very devil to eliminate, stealing food, spreading filth and disease and even attacking weak or sickly birds if the opportunity should arise. Late autumn and early winter is the time to be extra vigilant for signs of their moving in to establish a colony after spending a summer of plenty in the cornfields, and at the first indication of runs, holes and droppings, a campaign of elimination must be carried out immediately.

The ways of reducing the rat population are many, though a system of baited traps is a good standby all year round. A spring trap is the most effective and widely used, but must never be set in the open, both to comply with the law and for the sake of other wildlife. It should be covered with a long box tunnel to prevent accidental captures, and the entrance hole should be narrowed down to admit nothing larger than a rat, stoat or weasel. Luckily, rats can seldom resist exploring any hole or tunnel on their territory, particularly if a handful of wheat or pellets is sprinkled carefully beneath the trigger plate, and in this way all but the wariest can be taken. A piece of fish is probably the best bait of all for a rat, as indeed it is for many other types of vermin. To ensure all catches are taken cleanly, the size of the tunnel is most important, for it must be small enough to ensure anything entering the tunnel will pass directly over the trigger plate, though large enough not to interfere with the correct functioning of the trap, the tunnel being about 2 ft (60 cm) long overall. Set the trap halfway along the tunnel (Fig. 5) after scraping out a rough hollow to lower the trigger plate level with the ground. Conceal it carefully with a few strands of grass or dead leaves and set lightly so that it will spring at the slightest touch. The actual placing of the traps is also quite critical, for there are favoured spots where rats move which will catch time and again, and others where a rat is seldom taken. With trapping it is a matter of learning the signs. During the first three months of my waterfowl venture I kept a couple of tunnel traps working constantly, having noticed quite a few well-worn tracks beside the nearby river. During that period I took no less than 76 rats, having positioned one tunnel along the perimeter fence beside the river bank and the second directly opposite and adjacent to the open fields. That the rats obviously favoured the river bank was evidenced by the fact that out of the entire catch only three were from the field edge, the remaining 73 taken in the single trap beside the river. It goes without saying that any set trap must be inspected at least once daily, for however bad a pest has become, it must be dealt with as humanely as possible. The trap should also be sprung and reset periodically to prevent it seizing up. If a half-eaten duck is discovered, disturb it as little as possible and place the tunnel over it with a trap set at each entrance. The culprit

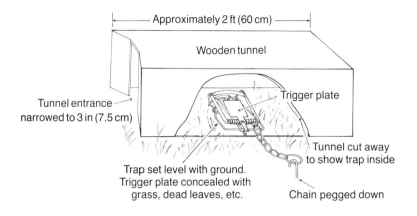

Approximately 2 ft (60 cm)

Wooden tunnel

Trigger plate

Tunnel entrance
narrowed to 3 in (7.5 cm)

Tunnel cut away
to show trap inside

Trap set level with ground.
Trigger plate concealed with
grass, dead leaves, etc.

Chain pegged down

Fig. 5   Spring trap and tunnel

will usually be taken when it eventually returns to clear up the remains.

The use of poison is probably the most effective method of rat control, but here again great care must be exercised in its handling and use. Poison stations must be well covered and access by anything larger than a rat restricted, for poison will kill a duck just as effectively, particularly as many of the modern types are made palatable by mixing with grain.

Another word of warning. Almost any brand of poison eventually loses its potency, the humble rat having an astonishing ability to build up an immunity against almost anything used exclusively over a long period of time. For this reason switch brands occasionally. One type that I used with lethal results for two seasons finally lost its effect, rats thereafter seeming to thrive on it and consuming the mixture wholesale without showing any reduction in numbers. When I switched poisons, the entire colony was wiped out in a matter of days.

All poisoned rats should be buried to prevent them being discovered by predatory birds, so that the poison's effect does not spread higher up the food chain. Use gloves and avoid handling the corpses whenever possible, for besides their other doubtful attributes, they can be carriers of the lethal Weil's disease, a deadly virus transmitted in the urine, which can enter a cut or graze on the skin with lethal consequences. In fact, avoid handling anything where rats are in evidence, for fear of contracting this most foul of diseases. Always use gloves when taking one from a trap as, even if killed instantly, a rat will often pass urine in its final throes, thereby possibly infecting the trap.

A last resort is to gas the rats in their burrows. A fine powder can be obtained which gives off a lethal gas when introduced to the burrows. It must therefore be treated with all due respect for its dangerous properties, used only on a dry, windless day and with a companion on hand in case anything should go wrong. Once all the entrance holes to a colony have been discovered and cleared, a spoon tied on a long stick is used to deposit a measure of powder into each hole, the entrance of which is blocked immediately with a sod of turf to prevent any rats from escaping. Properly used, and with no bolt holes left unblocked, gassing will destroy everything in the burrows.

Although far less common than the ubiquitous rat, the stoat is another enemy that must be kept in check at all costs, for should one happen to gain entry to an enclosure it will cause havoc among the resident fowl. It is capable of killing even the largest of ducks and not averse to robbing a nest of eggs. The stoat is an agile climber and will try to gain entry by scaling overhanging trees or wire netting if something attracts its attention, though all but the most determined can be deterred by an overhang of small-mesh wire incorporated into the perimeter fence as described in the section on fencing. Stoats will often be taken in tunnel traps along the wire as they travel around the pen looking for a way in as, like the rat, they seem unable to resist investigating any hole or tunnel encountered on their travels.

That stoats will steal eggs has been proved to me on two separate occasions. Twice I have encountered one in the act of rolling eggs along the road. The first was within yards of a known pheasant's nest along a roadside verge; and the second, early one morning on a quiet country lane, was rolling a large white egg along with its forepaws in a most competent manner. Stopping to examine it I found it was a hen's egg, and I was fully half a mile from the nearest hen-house. The stoat was possibly taking it back to its nest. Replacing the egg on the road I drove a few yards on before stopping to watch. The stoat immediately popped out of the hedge, found the egg and after a few cautious sniffs dribbled it skilfully along the road until eventually out of sight.

If a stoat is seen hunting and one has a gun handy, it is an easy matter to call it into the open. If you imitate the squeaking of a mouse, its natural wariness will be overcome by the prospect of an easy meal.

Though it does much good by way of killing countless small rats and mice, the weasel is little less of a menace to waterfowl than its larger cousin, even taking into account its diminutive size. There is still some confusion between the two animals, even among countrymen, but the most easily recognized way of telling them apart is by the tail, the stoat having a proportionately longer tail tipped with a tuft of black hair, whereas the weasel's tail is considerably shorter and of a plain brown throughout its length. The weasel is also much smaller, but despite its size is still capable of tackling all but the largest of ducks and poultry, possessing the ferocity and courage of an animal many times its size. Small ducks are no match for it, particularly the teal species which are at their most vulnerable at nesting time. Almost snake-like in their movements, weasels will take advantage of every available scrap of cover when moving around or hunting, the slimness of their lithe bodies enabling them to make use even of mole runs as a means of moving around unnoticed. Weasels can be taken in very lightly set tunnel traps or squeaked out of hiding like a stoat, and I have heard of them being unwittingly taken in mole traps on more than one occasion.

The best defence against a marauding fox is a sound perimeter fence, but on occasion a particular individual may become extremely persistent, attempting to dig its way under the fence or tearing at the wire in an effort to get in, and in such cases it may become necessary to deal with it rather more severely. Foxes are easily caught in steel snares, but must be dealt with quickly and humanely when caught, which means inspecting each snare at least once daily, preferably in the early morning, to avoid a captured animal suffering any unnecessary stress.

If one lives in an area where feral mink have become established, suitable precautions must be taken to guard against them, for the mink is a ruthless killer that will leave a trail of devastation in its wake should it have the misfortune to gain entry to a waterfowl pen. Not indigenous to Britain, from time to time mink have escaped from fur farms, quickly reverting to their wild state and easily earning a living from the land. The problem has been greatly exacerbated by well-meaning 'animal-lovers' who have deliberately released mink from captivity. These misguided souls should witness at first hand the wholesale destruction feral mink can cause, for they are certainly doing the animal world no favours, as can be seen in areas where mink establish themselves, the countryside quickly becomes devoid of much of its wildlife, including birds, small mammals, fish and amphibians. As mink live mainly along water courses and are excellent swimmers, waterfowl in particular suffer heavily from their depredations. Any mink living in the vicinity of a waterfowl collection will soon be tempted to investigate, but fortunately a good fence will keep them at bay; and the mink is easily trapped, both in cage and tunnel traps. The traps, set near water courses where mink footprints are in evidence on the mud, are best baited with a fish-head or a kipper, though anything with a strong fish smell will attract them. With a tunnel trap a larger version of the spring trap used for rats can be employed, which must be powerful enough to cope with a mink humanely.

Harmless though they may seem, moles can cause considerable damage in a waterfowl pen, quite apart from making an unholy mess of lawned areas by continually throwing up mounds of soil. The damp earth surrounding a pond always holds a lot of earthworms, which in turn attract the moles. Their tunnels, dug under the perimeter fence, can allow access to weasels, and they have an annoying habit of tunnelling directly beneath nesting birds with the result that the nest is often destroyed.

Moles capture their food – mainly earthworms – by excavating an extensive underground system of tunnels, every so often casting up molehills to dispose of the spoil. These 'runs', regularly patrolled, can be found by probing the ground between the molehills with a thin metal stake. Once the run is found, the mole can usually be taken easily with a scissor-type trap, with the jaws opened out to cover the run and the small trigger plate or tongue set in the centre

Mole caught in a scissor-type trap

of the tunnel to spring the trap as the mole noses it out of its path. To set the trap, excavate a short length of the tunnel just sufficient to allow the jaws to be inserted on each side of the run, clearing the tunnel of soil apart from a small mound in the centre in which the tongue can be concealed. Ensure that the trap can spring unhindered by soil or stones, and carefully re-cover the tunnel with turf, sprinkling lightly with soil to block out all the daylight so the runs shows no sign of being exposed. The handles, which remain above ground, will spring open to indicate when the trap has been sprung.

A mole can also be dug out while it roots close to the surface by creeping very quietly up to the working and then jumping on it as the mole pushes upwards, at the same time stamping the soil down firmly around it. With any luck the mole will be hemmed in long enough to allow it to be dug out, but always stamp the earth down tightly before digging or it will be long gone before you can reach it. They can even be killed with a shotgun if one approaches very cautiously to close range, but having done it several times I must admit resorting to such measures does seem rather drastic.

Mice are not really a problem provided all feedstuffs are stored in a mouse-proof hopper, but will gnaw holes in feedbags and contaminate food with their droppings if given the chance. They can be poisoned, but then will often die in the open where there is a great likelihood of a hunting owl or hawk picking them up and ingesting the poison. Traps are much safer, and mice are easily eliminated with a backbreaker trap baited with the customary lump of cheese.

Grey squirrels will plunder nests if given the chance, and there is hardly a fence that will keep a persistent squirrel out once it has acquired a taste for eggs. Poking out nearby dreys, shooting and cage-trapping, the latter baited with whole maize or wheat, will help overcome the problem.

Although few would wish them any harm, hedgehogs can become a nuisance in the breeding season, for they like nothing more than a nest of eggs, and with their covering of defensive spikes can even push an incubating bird from her nest to get at them. I have heard of more than one instance of a hedgehog gaining entry to a hen coop and killing and eating an incubating broody hen which was trapped inside and unable to retaliate against such a heavily armoured attacker. An enclosure is easily cleared of hedgehogs, and once they are removed outside the fence, will cause no further trouble. To capture them, dig a steep-sided hole about 18 in (45 cm) deep, undercutting the sides and baiting it up with something attractive. A rabbit paunch makes an

Damage a magpie can cause to a nest of eggs

ideal bait, but anything with an attractive smell, even tinned cat or dog food, will work just as well. Hedgehogs move around mainly during the hours of darkness, and in the morning Prickles will be found rolled tightly in a ball at the bottom of the hole unable to find a way out. Either fill in the hole after use, or continue to inspect it regularly, otherwise it could prove to be a deathtrap to other creatures.

Of all the winged vermin likely to cause problems in the spring, the carrion crow and magpie undoubtedly head the list, being ardent robbers of nests and killers of baby ducklings. Instead of haphazardly searching for a nest, these crafty birds will study the duck's movements from a suitable vantage point until its whereabouts is disclosed, and once successful will return time and again until all the eggs have been removed. The crow often carries off the eggs to a safe spot to consume them at its leisure, but the magpie, being unable to carry any but the smallest, usually eats them near the nest, leaving a scattering of egg shells to betray its evil deeds. Having discovered there are easy pickings to be had in a duck pen, both birds will make use of every opportunity to return for other nests and can prove extremely difficult to catch in the act, for both are wily, sharp-eyed, and will in due course become very crafty. An early-morning vigil with the gun can prove successful, but one needs to be in a concealed position well before first light. If a number are paying an early morning visit a more effective method is to construct a large wire-netting trap which can be baited to entice them inside, using either hens' eggs, white bread or possibly a dead rabbit or some other form of carrion. The cage trap, anything from 6–10 ft (1.8–3 m) square – the larger the better – is built with a series of narrowing wire-netting funnels leading inside, the end of each entrance just

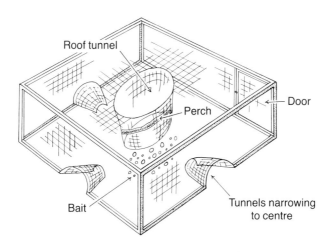

Fig. 6   Vermin cage trap

sufficiently wide to allow the bird to enter (Fig. 6). A vertical funnel is incorporated in the roof with a perch halfway-down, for birds will often perch on the roof of the trap and fail to find their way in at ground level. This funnel can be made quite large to allow easy access, as once inside few will find their way out again. Before setting, the trap needs pre-baiting for a week or so with one side left completely open to gain the birds confidence. Once they are taking the bait freely, the last side can be wired in completely to start catching. Such a trap will also take its toll of jays, jackdaws and rooks, all of which will plunder eggs and ducklings should the chance present itself, and in this way many a nest can be saved.

The odd rogue tawny owl can be troublesome during the nesting season if any young ducklings are allowed to be reared by their natural parents, but all species of owls are heavily protected by law and one should remember that the good they do in keeping down rats and mice far outweighs the occasional lapse when a duckling is taken. One owl's nest in a hollow oak overlooking the marshes near my duckponds often contains the odd wild duckling or two at hatching time, when the owls are feeding owlets, but these are probably taken on the journey from the nest to the safety of water, where they are far less vulnerable. As most ornamentals are hatched and reared well out of harm's way, the tawny owl poses little problem.

3

Surface-feeding Ducks

As ALREADY SUGGESTED, the complete beginner to waterfowl-keeping would be well advised to start with a few hardy species of surface-feeding or 'dabbling' ducks on which to gain experience before progressing to more delicate fowl. Being for the most part quite sturdy, many species are comparatively easy to maintain and breed and are therefore cheaper to buy. All the species described in the following pages fall into this group and can be kept in fine fettle on a basic diet of poultry-growers' pellets mixed fifty-fifty with wheat, apart from during the breeding season when they should be gradually weaned to layers' pellets a few weeks before the first eggs are expected. Titbits in the form of seeded lettuces, lawn trimmings and other greenstuff will be appreciated by the birds, and they will relish a load of water-weed dredged from a marsh drain or river, comsuming the lot after dibbling around for hours for the aquatic life always present in such luxuries. All surface-feeding ducks feed by dabbling around on the surface and upending in shallow water, though will supplement their diet by flighting to feed on fields of grain or other crops in season. A few, like the shoveler, have developed a spoon-like bill lined with delicate lamellae, for filter-feeding almost exclusively on the surface. As a general aid to identification, most surface-feeders float higher in the water than the diving ducks, with their tails raised, and their wing speculum is more brightly coloured than that of the latter, often with a vivid metallic sheen.

Apart from the shelduck, which can be quite aggressive during the breeding season, almost any of the following species can be kept together quite happily in a mixed collection, though hybridization between similar species can occur, particularly among the various blue-winged species and the promiscuous drakes of both mallard and pintail. Following a hurried display these latter two seldom bother with the niceties of an elaborate courtship and on occasion will pursue and attempt to rape almost anything which happens to take their fancy.

Scarcely needing an introduction, the Mallard (*Anas platyrhynchos*) is probably the best-known species of wild duck in the world, and certainly one of the most widespread throughout the northern hemisphere, where it is widely bred and reared both as an ornamental species and a sporting bird, vast numbers being released annually on flight ponds and lakes for shooting. Being easy to obtain, breed, rear and maintain, the lowly mallard has probably started off more participants in the waterfowl fancy than all the other species put together, and is also the stock bird from which most of the domestic breeds of duck originated, as can be verified by the two curly tail feathers of many domestic drakes, a trait inherited from their wild mallard ancestors.

Few people would need a detailed description to recognize the bird, but the wild drake is no less beautiful for its familiarity, and were it not so common, it would undoubtedly be much sought after by collectors. With a deep

Mallard (*Anas platyrhynchos*)

bottle-green head, white neck-ring, deep red-brown breast and generally grey body, the drake is not always welcome among a mixed collection of fowl, for it will hybridize freely with almost any species. For this reason, my ducks outnumber the drakes by about three to one, the surfeit of ladies helping to dampen their ardour. To date I have experienced very little trouble with interbreeding.

Mallard have a prolonged breeding season, often starting in late winter and running through to mid-summer, especially if the eggs are lost to predators or collected as soon as a clutch has been completed. In this way a single duck of mine produced a total of 101 eggs during one season a few years ago, though this can only be regarded as exceptional.

Besides keeping the drakes occupied, the high number of females provides me with plenty of foster-mothers to incubate otherwise difficult eggs (notably hooded mergansers) that hatch less well in an incubator than with a natural mother, for the mallard makes a very reliable broody once she is sitting and will put up with a great deal of disturbance before she will even think of deserting her nest.

In common with most of the northern-hemisphere fowl, in late summer the mallard drake moults into an 'eclipse' plumage, when it resembles the mottled brown duck in appearance apart from its yellow-green bill, whereas the beak of a duck is brown mottled with black. Its loss of gaudiness helps to conceal it better during the wing moult, when the bird is temporarily flightless as it simultaneously moults and regrows a completely new set of primary feathers. Most wild birds prefer to skulk around under cover of dense waterside vegetation until the primaries have grown sufficiently to allow flight.

The mallard will nest almost anywhere, amongst thick vegetation or high above the ground in tree holes, and will readily take to nest-boxes, baskets and tunnels if provided. One can often obtain a clutch of eggs for the asking, and a pair of adults is quite cheap. The ducklings, which hatch after 28 days of incubation, are very hardy and simplicity itself to rear, requiring only a starter diet of chick crumbs and later being weaned to pellets and wheat. I have seen mallard reared wholesale under appalling conditions that would make any normal ornamental curl up its toes, but they seem to thrive in almost any situation. Young mallard also tame quite readily and are the ideal bird for the complete beginner.

Far less common than the mallard, though almost as widespread, the Gadwall (*Anas strepera*) is a bird somewhat underrated as an ornamental species, its overall appearance seeming rather dull in comparison to other fowl, at least when seen from a distance. It is only upon closer inspection that one can fully appreciate the subtleties of its plumage, the drake appearing mainly grey but in fact intricately marked, his flanks delicately pencilled with fine black vermiculations in a most attractive pattern. The head is mottled grey brown with a slate-grey bill, the breast flecked with black and the belly white. The black tail coverts and white speculum are very visible in flight. Feet and legs are orange brown.

The female rather loosely resembles a small mallard duck, as does her voice, but her white speculum and noticeable orange bill with a dark culmen stripe distinguish the two quite easily. In the wild, gadwall consume a lot of plant material, so a regular supply of greenstuff will be appreciated. The birds prefer to site their nest, a shallow scrape lined with plant materials found in the vicinity, amongst dense vegetation, usually fairly close to water in which the eight to twelve creamy-white eggs are laid. The young hatch after about 25 days of incubation and the ducklings present few rearing difficulties.

There are three species of wigeon worldwide, the European, the American and the gorgeous Chiloe from South America, all of which are found in ornamental collections.

The European Wigeon (*Anas penelope*) is chiefly a winter visitor to the British Isles, though a smaller resident population remains in the north to breed. In winter a large influx of immigrants from the continent frequent our coastlines, estuaries and saltings, a few venturing further inland to large lakes and reservoirs. The drake has a rich chestnut head, darker around the eye and with a prominent yellow crown. His chest is pink brown, the belly white with grey flanks and back. The bold white wing-bar is most noticeable in flight, as are the white underparts terminating in black tail coverts. The speculum is green, the bill slate blue with a black tip, and the legs and feet grey brown.

The female is a generally mottled rufous brown, somewhat variable in tone, with white underparts and a duller bill. The drake has a most distinctive and attractive 'Whee-oo' whistle, the duck a growling purr.

The European mixes well with other fowl, though being mainly vegetarian will benefit from a good area of grazing or alternatively a regular supply of water-weed. On a good diet the birds are easy to maintain and breed. They built a nest of raked-up vegetation lined with down situated in thick natural cover, often among rushes or dense nettle-beds at some distance from water. The wigeon is a comparatively late nester and it is normally late spring or early summer before the female lays her six to ten creamy-buff eggs, noticeably pointed at one end. Incubation, by the female alone, takes 25 days and the ducklings, dark brown with reddish faces, are not difficult to rear. The species is also available in an attractive 'blonde' colour form.

The American Wigeon (*Anas americana*), as its name implies, is a native of North America. It is rather less common in collections, and an overall duller bird which lacks the grey flanks and back of its European counterpart, being for the most part a deep lilac brown. The head colour is basically cream speckled on the cheeks with black, which gives it the alternative name of Baldpate. There is a metallic-green eye-stripe running through to the nape of the neck. Females are similar to *Anas penelope*. To prevent interbreeding between the two species, they must be kept separately, but even then the drakes will occasionally crossbreed with other receptive females.

The Chiloe Wigeon (*Anas sibilatrix*) bears little resemblance to the rest of the family apart from its shrill three-note whistle. It is a charming bird, both in habits and appearance, and has the advantage of having no eclipse plumage, remaining in full colour throughout the year to provide a welcome splash of colour when most other fowl are drab and colourless by comparison. The sexes are similar, though the drake's plumage is much brighter when the two are seen side by side. A native of southern South America, the Chiloe has a white face, the rest of the head and neck being black, with a metallic-green patch on either side extending to the nape. Both sexes have a small white cheek-patch, brighter in the male. The breast is finely barred black and white, the upperparts black with white edges to the feathers and the underparts white to the black tail. Along the flanks of each bird is a bold wash of orange. The black speculum is sheened with green and the bill slate blue with a black nail.

Pairs are devoted to one another throughout the year, the male remaining on hand to guard his incubating mate and even protecting the young, almost unheard of behaviour in the duck world. Although they normally nest in thick cover, Chiloes will readily take to nest-boxes, baskets and even tunnels, laying six to ten cream-coloured eggs that take 25 days to hatch. The ducklings are very attractive, dark brown above, lighter below and with a warm-brown face. The birds will usually lay a replacement clutch if early eggs are gathered, and sometimes manage three broods in a season.

Another ideal species for the beginner is the Northern Pintail (*Anas acuta*), a hardy bird slightly smaller than the mallard and almost as easy to cater for. Easily bred, they will often lay three clutches in a year, though the males are not too selective in their choice of partners and will occasionally fight amongst themselves. Keeping two females for every

49

male can help overcome this problem to a certain extent.

The male is an elegant bird, his slim figure basically silver grey with chocolate head and elongated black central tail feathers. His long, slender neck and breast is white, with a white stripe extending upwards across the back of the face. The back is striped with long black and white scapulars, the tail coverts black with a light patch of buff around the ventral region. The beak is slate blue with a black culmen stripe. The female is a mottled grey brown with lighter underparts. The northern pintail is also available in blonde, silver and ginger colour phases.

Pintail will nest almost anywhere, making use of natural cover, nest-boxes or baskets where clutches of relatively small eggs, usually numbering six to nine, take 23 days to hatch. The ducklings rear easily on chick crumbs and can be weaned to small growers' pellets at about three weeks of age. Like most other ducklings they will enjoy greenery such as duckweed or finely chopped lettuce, and will welcome some finely chopped hard-boiled egg in the early stages.

The Bahama Pintail (*Anas bahamensis*) is a bird of charming looks and disposition from South America. A worthy addition to any collection, the sexes are similarly coloured and have the advantage of no eclipse plumage, thus remaining attractive all the year through. The plumage is a light red-brown, spotted darker on the breast and flanks and more striped on the back. The forehead, crown and nape are of the same colour, with sharply defined white cheeks and throat. The bill is blue grey with a base of prominent scarlet. The male is slightly brighter with a longer white tip to his tail. During display he bows forward to the female, elevating his rear end high above the water, and with parted wings curtseys to display the vivid-green speculum, at the same time giving a wheezy whistle. The birds are unobtrusive and get on well with other fowl, though can be shy breeders. The eggs, laid in a well-concealed nest among vegetation, are creamy brown and take 25 days to hatch. The ducklings are unmistakable, dark brown above and yellow below with bright yellow faces.

The aptly named Northern Shoveler (*Anas clypeata*) is so called for its wide spatula-like bill, with which it sifts the surface film of the water for minute animal and plant life with the row of delicate lamellae that line the edges of its bill. A pair of shovelers often follow one another when feeding, the following bird picking up food stirred up in the wake of the leader.

The drake has a deep bottle-green head sheened with blue, a mainly white body, chestnut flanks and a mantle of dark brown. The wing speculum is green, the forewing blue. The spoon-like bill is black, feet and legs bright orange. Females are a nondescript mottled brown, but easily identified from other species by the characteristic bill.

The drake's display consists of a rapid pumping up and down of the head, an action accompanied by a grunting 'Tok, Tok' to which the female responds in like manner. The nest is sited amongst thick cover, a shallow depression lined with local materials and down. The eggs, pale buff, are incubated for 24 days. Problems can occur with youngsters, which are quite delicate. Besides chick crumbs they need a quantity of live food such as mealworms or maggots, and an occasional treat of duckweed. Suitably catered for over the first few days they are not difficult to rear.

Of the remaining types of shoveler, the Argentine Red Shoveler (*Anas platalea*) is perhaps the most widely kept, a most attractive bird, the male having a pinky-buff head freckled with dark brown, somewhat heavier on the crown. Its breast and flanks are warm cinnamon spotted with black.

The long scapulars are black striped with white, the bill black, feet yellow brown and the iris white. A longer tail gives it a more streamlined appearance than others in the family. The female resembles the northern bird, though slightly paler overall, and, like the northern, both birds have the pale blue forewing and green speculum, though the Argentine has the advantage of no eclipse plumage.

No collection would be complete without the beautiful Mandarin (*Aix galericulata*), a bird that almost defies description, possessing all the colours of the rainbow blended together in a most attractive and striking pattern. Coming originally from eastern Asia, the mandarin has now established itself as a regular breeding species in some parts of Britain from a few captive escapees, though it is becoming rather scarce in its native lands. The drake in full colour has a distinctly oriental look, from the pink tip of his crimson bill to the brown-webbed orange feet, the most striking feature of all being the pair of enlarged orange tertiary feathers, one to each wing, which are carried permanently erect like a pair of miniature sails. His crown is purple running back to a metallic-green crest, and below the white face a shimmering orange beard creates a ruff around his neck. The breast is white, the flanks cinnamon behind a double slash of black and white, and the back, although dark, is sheened with a metallic blue. The neck is also dark and sheened with rich maroon.

The female provides a complete contrast, being a sleek, powdery grey-brown bird with lighter markings, a slight head crest, and a white stripe surrounding the eye that terminates at the nape of the neck. The drake loses his finery during the short eclipse moult, when he resembles the female apart from her slate-grey bill. It is during the mating display that the drake really shows off his finery, erecting his crest and puffing himself up ostentatiously as he struts around, bowing frequently to his mate and emitting a series of grunts and croaks answered by the female in a rather high-pitched wheeze. He is devoted to his partner, a fact recognized in ancient Chinese folklore, for a pair were often presented to newly-weds as a sign of fidelity. In fact the mandarin has never been known to hybridize, even in captivity, and will get on extremely well with any other birds.

A tunnel or box is preferred for nesting, for in the wild a natural tree cavity is usually chosen, often at a good height above the ground. The eight to twelve white eggs are incubated by the female alone for about 30 days. Being tree-hole nesters, the ducklings are renowned escapologists and the brooder needs to be covered during their first few days of life to prevent them from escaping.

Rivalling the mandarin in colour, the Carolina (*Aix sponsa*) is also called the wood duck. Unlike the mandarin, this bird is quite promiscuous and will occasionally hybridize with other fowl. The females of both species are similar, the carolina being somewhat darker overall with a green and purple tinge to her plumage and a bold white ring around the eye. The male is a splendid bird, his dark head sheened with greens and purples criss-crossed with slashes of white across the face and a wide throat-patch. The maroon breast is studded with flecks of white, the flanks are cinnamon edged with black and white leading to purple on the rump and a comparatively long tail. His back is almost black, again sheened with shades of purple, blue and green that shine iridescently in sunlight. There is a definite scarlet ring around the deep-red eye and the red bill, shading to pink with a black nail and culmen stripe, is edged with orange at the base. Feet are dull orange with dark brown webs.

When displaying, the male shows his crest to advantage by flicking his head sideways towards his mate, carrying out ritualistic preening movements with wings spread to display his colours. The birds will perch readily in trees and the natural nest site is a tree cavity or old woodpecker nest, but in captivity they will take to nest-boxes or tunnels cut into a sloping bank. The carolina is a prolific egg-producer. Given an early start, and if the eggs are collected as soon as a clutch is complete, the female will sometimes manage three clutches before her mate goes out of colour. The eggs, small, white and round, number from eight to fourteen, though less are laid in each successive clutch. Incubation takes 30 days. The ducklings can prove troublesome, at first devoting their energies to trying to escape by scaling the sides of the brooder, but once they are feeding well there are usually no further problems.

A close relative of the carolina, the Australian Wood Duck (*Chenonetta jubata*) is widely distributed throughout Australia and Tasmania. About shelduck size, the bird is often referred to as the Maned Goose, for the male is capable of erecting the feathers on the back of his head to create a tufted mane. Like a goose, it spends much of its time grazing and is seldom seen on the water, where it swims rather gracelessly. The plumage of the sexes differs noticeably, the male being a strikingly patterned bird with dark chocolate head, a mottled breast and light grey body, marked with black on the belly, back, ventral region and tail, with a prominent streak of black on the scapulars. His bill is black, legs and feet dark grey. Both sexes have black primaries, conspicuous white secondaries and a grey forewing terminating in black secondary coverts. The female has a much lighter head with a buff eye-stripe and superciliary band. Her body colour is less striking than the males, being grey brown mottled with black and white.

Despite its size the Australian mixes well with smaller fowl and seldom quarrels, taming quite readily. Chiefly a grazing bird, it prefers a grassy enclosure and is also fond of perching above the water. The wild bird chooses a tree hole for her nest, so an attempt should be made to provide something similar to induce them to breed or a large nest-box which should be situated off the ground. The eight to eleven creamy-white eggs hatch at 28 days. Fully-fledged juveniles resemble the female, but are generally lighter in colour and somewhat browner.

A conspicuous bird of coastline, estuary, salting and inland lake, the Shelduck (*Tadorna tadorna*) is unlikely to be confused with any other species. Its striking plumage, mainly black and white but with a broad band of bright chestnut around the forepart of the body, is prominent against any background. Its head is very dark green, almost black, the belly dark brown. The secondary feathers are sheened with greens and purples and the primaries are black. The sexes are similar, though the drake is larger and more colourful and has a large knob at the base of the upper mandible that swells visibly as he reaches breeding condition. The bill itself is bright red and the legs flesh-coloured. There is no eclipse plumage.

While on coastal saltwaters, the shelduck lives mainly on the small *Hydrobia* snail, which it discovers by sweeping its bill from side to side in the surface of the mud; but in captivity it will thrive on a basic diet of pellets and wheat if it has the opportunity to graze as well. Like all the shelduck family the males are aggressive and argumentative during the breeding season, often fighting amongst themselves and harassing other birds that venture too near their mate. If they are to be kept among other fowl, the birds need plenty of room or casualties can occur.

In the wild, shelduck use rabbit burrows and tree holes for their nests, and in captivity deep tunnels are usually favoured, though they will use nest-boxes provided there is a long entrance tunnel leading to the nesting chamber. The female usually lays a good clutch of anything up to 15 large creamy-white eggs that hatch at 28 days. The ducklings are possibly the prettiest of all young waterfowl, being like little pied black and white bundles of fluff when hatched. They will rear easily on a diet of chick crumbs switched to growers' pellets at about three weeks.

Also inclined towards aggression, though rather less so, is the Ruddy Shelduck (*Tadorna ferruginea*), a handsome bird, the attractive plumage of both sexes being a bright orange-brown with buff heads, the colour contrasting sharply with the black primaries and tail, dark grey legs and beak. There is no eclipse plumage but in spring the male assumes a black neck-collar. Like geese, the birds are good watchdogs but can be rather noisy, setting up with their almost goose-like calls the moment a stranger enters their enclosure. In the wild the birds prefer a tree hole often quite high above the ground for their nest, so a large nest-box or hollow log with a ramp leading up to it is most likely to be used. Clutches of eggs can be large, the white eggs taking 30 days to hatch. The youngsters resemble those of the common shelduck.

The largest member of the shelduck tribe is the Australian Shelduck (*Tadorna tadornoides*), kept in many collections both for its ornamental qualities and for its rather less aggressive nature when compared to other members of the family. The sexes are similar, the body a dark sooty-brown with rufous breast and black head sheened with metallic green. Both have a white neck-ring, the female differing by also having a white ring around the eye. Like all other shelduck, their tertiary feathers show as a prominent patch of chestnut and the forewing of both sexes is a contrasting white, very prominent in flight. The secondaries are sheened with green. Bill, feet and legs are black.

The Australian shelduck will welcome plenty of vegetable matter in its diet and, given the opportunity, will readily graze.

One could form a waterfowl collection consisting entirely of teal, for at present there are at least 18 species available from around the world, most of which are are easy to cater for and present few problems in their keeping. All are dabbling ducks, the smallest of the waterfowl. Many are quite colourful. In most cases the species of the northern hemisphere are sexually dimorphic, the drakes in full breeding dress being far more colourful than their soberly dressed partners, though resembling them in eclipse moult, while those from the southern hemisphere are almost identical in plumage except to the practised eye, the males generally being slightly larger and brighter. Most of the latter group stay in colour all the year through, appearing only slightly duller during the moult. The following are among the better known and most widely kept species.

One of the most attractive is the European Green-winged Teal (*Anas crecca crecca*), so-called for its vivid-green speculum, a bird which ranges widely across the northern hemisphere and is second only to the mallard in terms of population. A jewel of a bird in full dress, the drake has a chestnut-red head with a broad green stripe edged in yellow from the eye to the nape, a light-grey body pencilled darker on the flanks, and a spotted breast. The undertail coverts are yellow edged with black. The female is dull brown mottled with darker markings.

At breeding time the drakes are quite vocal, setting up a musical piping as they display with their tails up, bobbing towards the female of their choice. Green-winged teal should be kept as a single pair, or if in a small colony the ducks should outnumber the drakes, for the drakes will pursue and harass the females to mate and will sometimes gang up to achieve this end. The drakes are very persistent and I have even known instances of females becoming waterlogged and drowning. With too many males fights will also break out.

The nest, usually built in low, dense vegetation, is sometimes placed at quite a distance from water. The eight to ten eggs are light olive-brown in colour and take 23 days to hatch. The youngsters, although quite tiny when first hatched, usually rear well once encouraged to feed.

Its close relative, the American Green-winged Teal (*Anas crecca carolinensis*), is very similar though it lacks the yellow edging to the green eye-stripe and has a vertical white stripe running from shoulder to breast.

The Garganey Teal (*Anas querquedula*) visits Britain in small numbers during the summer months, and keeps mainly in the south-east. Hence it is sometimes referred to as summer teal. The drake has a dark-brown crown, mottled reddish-brown cheeks and breast, grey flanks finely vermiculated with black, and a bold white superciliary stripe running from the front of the eye to the nape. The slate-blue scapulars are striped with black and white. Garganey are cheerful, lively little birds, always busy. They tame easily, and as soon as the males acquire their breeding plumage are quite vocal as they display to their mates by producing a curious clicking sound, quite unlike any other duck, which gives the bird yet another alternative name of cricket teal. One drawback is that the males rarely begin to colour much before late winter, and it is often early spring before their full breeding dress can be appreciated, being in eclipse for well over half the year, when they resemble the mottled brown females apart from the grey forewing of the drake.

Garganey prefer to nest among low cover, often well away from water. The eight to ten buff-coloured eggs are laid during the spring (late April or May in the UK). Incubation lasts 23 days, and if a clutch is collected it will often be replaced as late as June.

The Blue-winged Teal (*Anas discors*) is common in collections and mixes well with other fowl, though can be somewhat shy and secretive in its behaviour. The breeding male has a slate-blue head with black crown, forehead and throat, and a bold crescent-shaped blaze of white in front of the eye. The body is light brown thickly mottled with black, the upperparts darker brown and tail coverts black, with a white patch on either side of the rump. He has a bright sky-blue forewing conspicuous in flight. The female is mottled brown with a pale patch at the base of the bill. Her blue forewing is noticeably duller than the males. In eclipse both sexes are similar but the drake has a darker crown.

Blue-winged teal are quite likely to hybridize with other blue-winged species, such as cinnamon, garganey or even shoveler, if given the chance, but if all goes well a pair can produce a lot of youngsters in a season, three, or even four clutches are not unheard of. The attractive ducklings, very tiny at first, are bright yellow with darker markings and hatch at 23 days. If they survive the first few days, progress is rapid and they are fully fledged at six weeks of age.

Chosen more for its gentle nature than its ornamental qualities, the Marbled Teal (*Anas angustirostris*) is nevertheless quite attractive and a charming bird of kindly disposition. Tame and confiding, it will mix readily with other fowl without fuss. The sexes are identical apart from the longer and whiter tail of the drake, the main body colour being pale grey tinged with brown, spotted and 'marbled' with light grey. There is a dark patch around the eye extending to the nape. Chiefly resident resident around the Mediterranean, the marbled is unusual among northern-hemisphere birds for it has no eclipse plumage. The male displays by withdrawing his head into his shoulders, erecting his crest and jerking his head upwards and giving a rather rusty squeak. Marbled can be prolific breeders. One pair I had made four different nests during one season, producing a total of 44 eggs. Their eggs are small, white and hatch after 26 days. The ducklings are powder grey above and lighter beneath. Even though they are tiny they are quite hardy and present few problems in rearing. However, possibly due to prolonged interbreeding, many captive strains have developed a weakness in the legs and when caught up need handling very carefully.

A beautiful little duck from central South America, the Ringed Teal (*Callonetta leucophrys*) adds colour to a collection throughout the year, for although sexually dimorphic the drake has no eclipse plumage. His buff face is bordered by a black crown and nape, the pink breast is spotted with black and the flanks are grey. The mantle is fawny grey, with bright chestnut-red scapulars, while the tail coverts are black with a conspicuous white spot on either side of the rump. The female is brown, her flanks barred lightly with white, with white throat, face markings and a broad superciliary stripe. Both sexes have pink legs and feet, grey bills, green specula and noticeable white wing-patches. Pairs are devoted to one another, seldom seen apart and quickly becoming tame. The mating display is little more than a low whistling by the drake as he swims around his partner. Ringed teal are agile climbers and love to perch on a log above the water, favouring tree holes for the nest, though nest-boxes will be used for the six to eight white eggs that take 23 days to hatch.

Originating from central and southern Africa, the Cape Teal (*Anas capensis*) is a popular collectors' bird. It is soberly though attractively marked, the pale pearl-grey body spotted with dark brown, and the head finely speckled with the same colour. The scapular feathers are dark brown edged with fawn, the bill pink with black at the base and along the leading edge of the upper mandible. Male and female are similarly marked, though the drake's bill becomes noticeably brighter in the breeding season, an extended period that can last through until late summer. The nest is made on the ground among cover or under a bush and lined copiously with down. Six to nine creamy-brown eggs are laid, that hatch at 23 days. Although normally of an agreeable nature, during the breeding season the birds can become rather territorial, defending a chosen area from trespassing fowl.

Natives of South America, the sexes of the Versicolor or Silver Teal (*Anas versicolor*) are virtually identical, the drake being in fact a little brighter and having a larger salmon flash on each side of the bill, which is blue with a black culmen stripe and nail. It is one of the prettiest teal, with a dark, almost black crown and nape, a cream-coloured face, a fawn-mottled breast and heavily barred grey flanks. Despite settling down and taming quite easily, versicolor breed only sporadically, but when successful the bird lays eight to ten eggs among natural cover or occasionally in a nest-box. The eggs are a light pinky-buff and take 25 days to hatch. The few eggs I have collected have often been deposited among other clutches, usually spread among more than one nest, though this occasionally happens with several other species at favoured nesting sites.

In many ways resembling the versicolor, the tiny

Hottentot Teal (*Anas punctata*) from southern and central Africa and Madagascar is one of the smallest ducks in the world. The sexes are similarly coloured, the male being more brightly decked out. His buff cheeks have a dusky patch running to the nape topped with a low crown of blackish-brown. The main body colour is dull cinnamon with dark-brown markings on the breast, the dark-brown feathers of the mantle bordered with cinnamon and the scapulars edged with light buff. The black wings are sheened with bottle green, the speculum is metallic green with a black subterminal border and white tips to the feathers. The beak is a light grey-blue with a black culmen stripe. There is no eclipse plumage.

Although the Hottentot will do well on a basic ration of pellets and wheat it will greatly appreciate a good proportion of vegetable matter, water-weed, and so on in its diet. Not the easiest of birds to breed in captivity, the nest, which is lined copiously with blackish down, is usually placed on the ground amongst vegetation close to the water's edge. The clutch size varies from five to eight small creamy-buff eggs, which have a short incubation period of some 20–22 days.

As its name suggests, the Chilean Teal (*Anas flavirostris*) is a native of South America, particularly Chile. Of a very gentle disposition, the bird is perhaps more prized for its absolute tameness than its decorative qualities. The latter, nevertheless, are a pleasing mixture of various shades of brown, greyer on the underside and richer on the back. The head is finely speckled with dark brown and the body attractively spotted and mottled with black. The metallic-green speculum is edged with buff and tipped white, the bill yellow topped with black, which gives it the alternative name of Yellow-billed Teal.

Despite their lack of colour, a pair of Chileans make a

pleasant contrast among more highly ornamental fowl and will breed readily. Using tree holes in their natural state, Chilean teal will normally make use of a nest-box but will also use a basket or make their nest among thick vegetation. The buff-coloured eggs hatch at 26 days of incubation. The nesting season can be prolonged from early spring until well into summer, with replacement clutches common if earlier eggs are gathered. The young are very tiny and require a little live food added to their diet, but otherwise present little difficulty.

A bird from the coastal wetlands of southern Australia, the Chestnut-breasted Teal (*Anas castanea*) is unusual among southern-hemisphere teal in being sexually dimorphic, the male having a much more colourful dress than his partner when he is in full breeding condition, though resembling her when in the eclipse moult. The drake has a dark-green head with a bright-red eye, a chestnut breast extending along the flanks which is separated from the black tail coverts by a prominent white ventral patch. His flanks are heavily spotted with dark brown. The upperparts are dark brown, almost black, with chestnut-buff edging to the feathers. The female is a general dark brown with buff feather fringes and a pale throat and cheeks. Her iris is duller than the male's. Bills and feet of both sexes are grey.

Although not easy to breed, the birds will mix happily among other fowl and, when successful, clutches can be large, anything up to 15 eggs being produced in a single brood. The nest is sometimes sited amongst vegetation on the ground, but the birds will also utilize hollow logs or nest-boxes, where they incubate their clutch of cream-coloured eggs for about 27 days.

About wigeon size, the Falcated Teal (*Anas falcata*) is one of the largest teal, and certainly among the most ornamental.

The female is a warm mottled brown with paler throat, a slight head crest and green and black speculum, but the male is quite stunning, his crown and cheeks a brilliant metallic bronze that changes to glossy purple in the rays of the sun. The central area of the head from the eye to the nape is a mane of rich glossy green. The throat is pure white above a breast of closely marked black and white, the flanks and mantle grey, vermiculated finely with black. An outstanding feature is his elongated, sharply curved tertiary feathers, black with white edges and a dark blue sheen, that hang on either side of the tail over a patch of buff bordered with black around the ventral region. The birds are somewhat shy in captivity and seldom become really tame, but under the right conditions can be persuaded to breed. Thick vegetation beside water is their first choice for a nesting site, but nest-boxes are occasionally used, sometimes well into summer. Six to ten creamy-buff eggs are usually laid.

The drake Baikal or Formosa Teal (*Anas formosa*) is an exquisite bird, its intricately patterned head distinguishing it from any other species. Slightly larger, though in some ways similar to the common green-winged teal, the face is a creamy buff, topped with a black crown edged with white and a black line extending from the eye to the black throat. A crescent-shaped metallic-green slash runs from the eye to the back of the head to the black nape. The slate-grey flanks are separated from a breast of pinkish brown by a vertical white shoulder stripe, and the ornamental elongated scapular feathers are striped rufous, black and white. The black tail coverts are quite noticeable. The female is a mottled brown bird with darker crown, white belly and a circular white spot at the base of the bill. Juveniles resemble her before their moult into adult plumage, but lack the white facial spots.

European Wigeon (*Anas penelope*)

Chiloe Wigeon (*Anas sibilatrix*)

Northern Pintail (*Anas acuta*)

Bahama Pintail (*Anas bahamensis*)

Northern Shoveler (*Anas clypeata*)

Argentine Red Shoveler (*Anas platalea*)

Mandarin (*Aix galericulata*)

Carolina (*Aix sponsa*)

Shelduck (*Tadorna tadorna*)

Garganey Teal (*Anas querquedula*)

Blue-winged Teal (*Anas discors*)

Ringed Teal (*Callonetta leucophrys*)

Versicolor or Silver Teal (*Anas versicolor*)

Chilean Teal (*Anas flavirostris*)

Falcated Teal (*Anas falcata*)

Cinnamon Teal (*Anas cyanoptera*)

Fulvous and White-faced Treeducks
(*Dendrocygna bicolor* and *Dendrocygna viduata*)

European Pochard (*Aythya ferina*)

Red-crested Pochard (*Netta rufina*)

European Greater Scaup (*Aythya marila marila*)

European Goldeneye (*Bucephala clangula*)

European Eider Duck (*Somateria mollissima*)

Long-tailed Duck (*Clangula hyemalis*)

Hooded Merganser (*Mergus cucullatus*)

North American Ruddy Duck (*Oxyura jamaicensis*)

European White-fronted Goose (*Anser albifrons*)

Greater Snow Goose (*Anser caerulescens atlanticus*)

Bar-headed Goose (*Anser indicus*)

Emperor Goose (*Anser canagicus*)

Red-breasted Goose (*Branta ruficollis*)

Hawaiian Goose or Ne-ne (*Branta sandvicensis*)

Black Swan (*Cygnus atratus*)

Distributed widely across eastern Asia, the Baikal breeds in central and eastern Siberia and winters mainly in Japan, where for some reason numbers have dropped considerably of late. It is still comparatively rare in collections, being of a shy and retiring nature and proving very difficult to propagate. Few have been bred and hand-reared, the main captive stock originating as wild-caught specimens, and as such the birds are expensive to buy. The nest is usually placed among light vegetation beside water, where an average of eight or nine pale-olive eggs are incubated for 24 days.

Although a teal-sized bird, the Cinnamon Teal (*Anas cyanoptera*) is in fact more closely related to the shovelers, which it resembles in its habits and in the long, black, slightly spatula-like bill that it uses for filtering mud and water through well-developed lamellae in order to extract the aquatic vegetation, seeds and minute animal life. The little drake is a gorgeous chestnut red, darker on the crown and with long black scapulars streaked with buff. The undertail coverts are black, feet and legs yellowish orange. Both sexes have a bright-blue forewing, the female otherwise being a mottled brown bird closely resembling the blue-winged teal. For this reason the two species should be kept apart to avoid hybridization occurring.

Normally preferring thick cover close to water, the cinnamon will occasionally make use of nesting baskets and boxes in which to lay its eight to ten pale buff eggs, which take 25 days to hatch. The youngsters can be quite delicate and fickle feeders, but a few mealworms and finely chopped greenstuff added to the dish of crumbs will soon encourage them to feed.

Resurrected from the very edge of extinction, the Laysan Teal (*Anas laysanensis*) now thrives in captivity and its wild population on the remote Pacific Laysan Island has been restored to a few hundred birds, possibly the maximum the tiny volcanic island can support, though numbers still continue to fluctuate. By 1912 only seven birds remained and rumour has it that by 1930 the numbers were even lower.

It is a soberly coloured though interesting bird, and interbreeding in captivity has produced several colour variations − blues, buffs and blacks. However, the original wild bird suggests a rather small, dark-plumaged mallard duck, with sooty-coloured head and an irregular patch of white around the eye. It has a grey-green bill and orange legs and feet.

Laysan teal tame readily and are prolific breeders, though clutches of greenish-white eggs are often small. Incubation takes approximately 28 days. As there is now a healthy stock of birds in captivity, the island can be replenished should some natural disaster again befall the wild population, thus ensuring the future survival of the species.

Of the eleven species that comprise the order Dendrocygnini, the treeducks, the two most commonly kept species in captivity are the White-faced Treeduck (*Dendrocygna viduata*) and the Fulvous Treeduck (*Dendrocygna bicolor*). Also called the whistling ducks for their most un-ducklike calls, the whole family are rather ungainly creatures and quite comical looking, all having proportionately large feet, long legs and a distinctly upright posture. The two species described are delightful birds to keep and quite easy to maintain, though breeding success is somewhat sporadic.

The white-faced is the most visually striking of the group, with a conspicuous white face, black nape, breast and underparts, a rich-chestnut neck and back, and flanks barred an attractive black and white. Found in both South America and Africa, it is a highly gregarious species, sometimes forming huge flocks in the wild. In a collection the pairs remain very close, keeping in touch with one another by an urgent, exaggerated wolf-whistling when separated and displaying their affection for one another by indulging in mutual preening.

The fulvous is no less pleasing, being of a rich fulvous colour over most of the body and head, with upperparts of dark brown barred with buff. The tail coverts are a deep creamy white, as are the large ornamental flank feathers which envelop the wing. Beak and legs are slate grey. Its whistle is rather less musical than the white-faced, being more like a rusty scream.

Both species are agreeable with other fowl and in a short while can become tame and confiding, their frequent whistlings adding an extra touch of charm to any collection.

4

Diving Ducks

Wɪᴛʜ ᴀ ꜰᴇw ᴇxᴄᴇᴘᴛɪᴏɴꜱ , the diving ducks need rather more specialized conditions than the surface-feeders if they are to thrive, the most important of all being a good area of reasonably deep water. With legs set well back on the body for easy underwater propulsion, the divers are rather ungainly on dry land. They prefer to spend most of their time on water, which needs to be, at the very least, 2 ft (60 cm) deep to allow them to swim and dive naturally, and should ideally contain a rich supply of aquatic animal and vegetable life with which they can enrich their diet in order to breed successfully.

Adapted to a more aquatic life than the surface-feeders, all diving ducks have relatively large feet with legs set well back on the body, small wings, and a more streamlined appearance. Most find it necessary to patter across the water to attain sufficient speed to become airborne, compared with the instant vertical take-off achieved by most surface-feeding species. Divers swim low in the water, their tails held parallel with or actually resting on the surface. Different species have developed specialized beaks to suit their feeding techniques; for example, the eider, which uses its strong and heavy bill

Tufted Duck (*Aythya fuligula*)

for dislodging molluscs from the sea floor, and the sawbill (merganser) tribe, which has evolved long, slender and serrated bills equipped with a hooked nail for catching and holding struggling fish.

The Tufted Duck (*Aythya fuligula*) is a familiar sight bobbing around on inland lakes, gravel pits, reservoirs and even town parks; in fact, almost anywhere there is a suitable area of open fresh water. It is seldom found on the sea except in really severe weather conditions when many of the inland waters have frozen over. Extremely buoyant, it dives well for a variety of plant and animal life, chiefly molluscs and insect larvae.

The little drake is unmistakable, being mainly black sheened lightly with greens and purples, but with pure white flanks that are conspicuous at a distance. He has a long black tuft hanging from the nape, and a vivid-yellow eye. The bill is slate blue with a black nail, the legs and feet slate blue with darker webs. The female is mainly dark brown, with paler flanks and a light belly. She also possesses a head tuft, though smaller than the drake's, and a white patch at the base of the bill which varies in intensity. In typical rapid flight both sexes display a white wing-bar.

Although quite easily kept in confinement, the birds need a regular supply of natural live food and aquatic weed if they are to be induced to breed. The nest is usually built on an island, or at the water's edge among thick bankside vegetation, in early summer (mid-May and June in the UK). The eggs, olive green varying to khaki brown, are very large for the size of the bird, as many as 14 being laid in a single clutch. Incubation takes about 25 days by the female alone, and the sooty-black ducklings, once feeding, are easily reared. In the wild, young families suffer heavy losses to pike, providing a vulnerable and tempting target as they bob around on the open water or imitate their parents by diving for food.

The Ring-necked Duck (*Aythya collaris*) is the North American equivalent of the tufted duck, which it quite closely resembles. It is confused in its native home with both greater and lesser scaup, though its darker back, and grey instead of white wing-bars, distinguish the species at a distance.

The male is very much like the tufted, though greyer on the flanks and lacking the long nape tuft of the latter. Instead, it has a short tuft on the top of the head which gives the appearance of a high crown. The beak is grey, bordered around the base and near the black tip with pale grey. The narrow chestnut collar which gives it its name is not prominent. The female has a much lighter head than the tufted, particularly around the base of the bill, with the addition of a pale ring around the eye and a definite eye-stripe.

In the wild the ring-necked is predominantly a vegetarian, and as such will enjoy a diet containing plenty of greenstuff. Water-weed and aquatic plants are ideal. The birds are not easily bred in confinement unless conditions are to their liking. The six to eleven eggs, laid near water, vary from pale buff to deep olive brown, and take about 28 days to hatch.

Another diving duck commonly found on inland waters over much of the British Isles is the European Pochard (*Aythya ferina*), the drake easily recognized by his bright chestnut-red head, black breast and tail coverts, and light grey body. He has a bright-red eye whereas the female's is brown. Feet and legs are similar in both sexes, grey with darker webs, as are their bills, almost black and both having a pale-grey central band. There the similarities end, the female being an inconspicuous blend of warm browns and greys.

Like the tufted duck, the pochard much prefers open fresh water. In fact, the two species are often found together in winter when they form gregarious flocks before the spring dispersal to breeding areas. Pochard breed quite well in captivity, though being more vegetarian will benefit from plenty of greenstuff added to the basic diet of pellets and wheat. During courtship the male stretches his neck forward parallel to the water, inflates his neck and utters a soft 'Yeeeow' to the female. The six to ten eggs are often laid in early spring in a nesting scrape among dense waterside vegetation, or even on a raft of floating weed, the female adding regularly to the construction until she is perched on a mound of vegetation. The eggs are a rich olive brown, quite large, and if collected early will usually be replaced by another clutch. The incubation period is roughly 26 days. Pochard ducklings are dark brown blotched with lighter markings, and have light faces. Hand-reared they will often become tame enough to take food from the hand.

The pochard has two American cousins, both similarly coloured, though there are noticeable differences between the species.

Still quite scarce in collections, though no less attractive, the American Redhead (*Aythya americana*) has a more rounded head and glaring yellow eye. Its upperparts and flanks are darker grey and the head of the drake is redder in colour than the pochard. The female is brown, with a light face and brown eyes.

The largest of the trio, the Canvasback (*Aythya valisineria*) can be distinguished, apart from its size, by the long bill, flat forehead, and rather elongated, wedge-shaped head which in many ways resembles the head of an eider duck. The bill is black, the head noticeably darker than the pochard and it has a very light body colour, almost off-white instead of grey. Although reputedly difficult to breed, my own pair regularly produces two clutches of fertile eggs each season, six to ten olive-grey eggs that are usually found in the same nesting basket year after year. When displaying, the drake inflates his throat visibly, throws back his head and emits a curious 'Crooo' to incite the female, who responds by lying flat on the water in readiness for mating. Being similar to the pochard, the female is easily told apart by her long black bill and browner plumage. The young are no trouble to rear, though rather ungainly off the water as they have an upright stance and rather large feet.

Not strictly a diving duck, the Red-crested Pochard (*Netta rufina*) prefers more of a vegetarian diet, with some insects, larvae and other aquatic life also occasionally taken. It seldom dives, getting most of its food by dibbling and up-ending for submerged vegetation in shallow waters.

The drake is the most colourful of the pochard tribe, with a velvety vermilion head, black breast and underparts lightly sheened with green, and conspicuous off-white flanks. The upperparts are dusky brown with a white stripe running vertically across each shoulder. The waxy bill and eyes are bright red, the legs and feet orange brown with dark webs. The female is dull by comparison, mainly brown with a dark crown and lighter face and breast. Her bill, eyes and legs are grey brown. During his eclipse moult the drake resembles the duck, though his bill retains some semblance of redness and the red eyes readily distinguish the sexes. A blonde form of the red-crested pochard has been bred in captivity, though pairs are more expensive to buy than those of the usual colour.

Widely though spasmodically distributed throughout Europe and as far east as India, an occasional visitor has been known to turn up in Britain, though not on a regular basis.

The red-crested is a good diving duck for the beginner, being quite hardy and seeming to do well almost anywhere, besides breeding quite prolifically in captivity. During the drake's display he carries out various head and neck movements, which include bill-dipping and letting out a wheezing call while erecting his crown feathers, which are puffed up to give him a top-heavy appearance. The bird can be an early nester, with the first clutch often completed by the beginning of spring (late March in the UK). In a good season three broods are not uncommon. In the wild, the red-crested usually sites its nest among thick vegetation, but captive birds will nest almost anywhere, utilizing nest-boxes and baskets as well as natural cover near the water. The eggs are pale buff, there are usually six to twelve, and they are incubated for about 26 days, the drake sometimes disclosing the whereabouts of the nest by standing guard within a few feet of his brooding mate. The young are not difficult to hand-rear but can be rather prone to feather-picking at quite an early age if kept in a confined space with nothing to occupy their minds. One way of alleviating the boredom is to supply them with plenty of greenstuff to peck at, and to remove them from the brooder to an outside pen as soon as possible, where they can move about and graze freely.

Like the red-crested pochard, the Rosybill (*Netta peposaca*) is another semi-diving duck of the pochard family. Living on South American fresh waters, the drake is a mainly black bird, highly glossed with greens and purples to the head and neck. The flanks are greyish, and undertail coverts white. An unusual feature is the large red knob at the base of the upper mandible, which extends well up the forehead. The actual bill shades to pink towards the black nail. The eye is red, deeper in the breeding season. There is no eclipse plumage.

The duck is a dull brown, darker above and also with white under the tail, and patches of buff-white on the throat and around the base of the bill. Her face is freckled with patches of the same colour and there is a definite ring around the eye.

Rosybills are not easily bred, but when induced to do so the female places her grey-green eggs in a typically pochard-like untidy accumulation of raked-up vegetation among thick cover beside the water. The eggs hatch at about 24 days.

Of the four species of white-eye, the most frequently kept is the Common White-eye or Ferruginous Duck (*Aythya nyroca*) found in the wild in eastern and southern Europe and as far east as India. It is a rare visitor to Britain. A relative of the tufted duck, although not so striking, the white-eye is an attractive rich mahogany brown on the head, breast and flanks, with dark-brown upperparts sheened with green, the belly white, shading to grey on the lower abdomen. It has conspicuous white undertail coverts and a white spot at the base of the lower mandible. Its bill, legs and feet are grey. The duck is similarly coloured, though duller, browner and lacking the white iris of her mate. She is easily confused with the female tufted, apart from the white undertail coverts which are noticeable at a distance.

The white-eye is basically a freshwater vegetarian living mainly on plants and seeds, but small aquatic life and grain are also taken. The birds will do well on a mixture of pellets and wheat, and will greatly appreciate a quantity of greenstuff. The nest is a woven construction of reeds and sedges lined with down, usually well concealed amongst waterside vegetation. The eggs, pale buff and varying from six to ten, take about 26 days to hatch.

Although related to and sometimes confused with the tufted duck at a distance, the European Greater Scaup (*Aythya marila marila*) is in fact a much larger and more colourful bird when observed at close quarters. Breeding in the northern tundra, the scaup visits the British Isles in great numbers during the winter months, remaining mainly at sea though almost always preferring to nest near fresh water.

The drake has a black breast and head, the latter sheened with deep bottle-green. The belly and flanks are white and the back pale grey pencilled with fine black vermiculations. The tail and coverts are black, the feet grey and both sexes have a very broad slate-blue bill, brighter in the male, tipped with a curved black nail. The eyes of both sexes are bold yellow. The female differs from the tufted in having greyer flanks and back, and a much broader collar around the base of the bill.

Scaup dive for molluscs, chiefly cockles and mussels, while on the sea, though during their summer stay on fresh water they take a quantity of greenstuff and seeds. Being adaptable, they are one of the most suitable saltwater ducks for the collector, as provided they have access to natural water they seem to thrive on a basic diet of pellets, though should not be allowed too much grain.

Breeding reasonably easily in confinement, the duck constructs a huge mound of wilting vegetation among thick cover sometime in early summer, the nest resembling a pochard's in which anything from six to twelve deep olive-green eggs are laid, like many of the diving ducks quite large for the bird's size. Incubation takes from 26 to 28 days. The youngsters also resemble tufted, though somewhat lighter in colour and with noticeably broader bills. They are easily reared on chick crumbs, but can be given a few mealworms as a treat. Like most young diving ducks they appear rather clumsy off the water and can experience problems with their feet. They should be given a shallow water tray on which to swim and bathe at an early age, and be put out on a pond as soon as fully fledged.

Two other species of scaup are kept in captivity, though rather less commonly. The New Zealand Scaup (*Aythya novae seelandiae*), or black teal, drake is glossy black sheened with greens and purples and the female resembles the greater, with the typical broad white band around the base of the bill. Both sexes have the black nail at the tip of the bill and in common with all other members of the family, bright-yellow eyes.

An inhabitant of North America, where it is the commonest of the diving ducks, the Lesser Scaup (*Aythya affinis*) is almost a smaller version of the greater, though it has a rather more purple sheen to the head.

Not easy to cater for unless one has a good area of natural water, under the right conditions the three species of goldeneye are certainly worth the effort, not least of all for their intricate courtship displays, which commence in early autumn as soon as the birds moult from the drab tones of the summer eclipse into full breeding attire. They are among the most attractive of the diving ducks, requiring at least 2 ft (60 cm) of water in which to manoeuvre and this should ideally contain a good supply of natural animal food essential to their diet. Biscuit meal containing meat can be given as a substitute, but in the wild the birds take a high-protein diet including shrimps and other crustaceans, mussels, small crabs, tadpoles and various insect larvae, besides a quantity of greenstuff.

Commonest of the trio is the European Goldeneye, (*Bucephala clangula*), the drake being a most striking bird with an almost black head richly sheened with bottle green and a prominent white, almost circular spot below and just forward of the bright-yellow eye. The breast is pure white, the back black, with black and white striped scapulars covering the wings at rest. The feet and legs are orange brown and the bill is black. The female is a much duller bird, though very attractive, with grey upperparts, white wing-bars and a white belly. She has a rich chocolate head, though lacks the white facial spot, having instead an indistinct white neck-ring. Her black bill is tipped with orange. Being diving ducks the legs are set well back on the body, giving the goldeneye a rather ungainly, upright shuffling stance when on dry land, but in their true element the birds are graceful swimmers.

Goldeneye have an elaborate and involved courtship display which is worth describing in detail. The drake begins by repeatedly going through his paces in what is called the 'head-throw' display, lowering his head across his back and kicking backwards as he calls to attract the attention of the female. She responds vocally and soon settles flat on the water in readiness for mating. After several such bouts of head-throwing the male, rolling over on one side, circles the female on the water, first one way and then the other, switching direction as each uppermost wing is extended in turn while he sails around her. Immediately prior to mating the male blows into the water half a dozen times, swivels his head to peck his scapulars and then mates, holding the female almost beneath the water. Following a successful mating the male powers away with neck stretched erect, calling proudly as he goes, after which both birds splash and preen vigorously. After a short interval the whole complicated rigmarole begins again. The whole process can take several minutes and is a delight to watch.

The goldeneye prefers a natural tree cavity in which to nest, and in the wild will often take over the disused nest of a black woodpecker. Nest-boxes are readily taken to in captivity, the bird sometimes laying a good clutch of nine or ten beautiful deep green eggs, quite large for her size. The ducklings are very beautiful, almost black with lighter underparts and snow-white cheeks. A clutch of day-olds is a most endearing sight although, being far from easy to propagate, the birds are quite expensive.

The American Goldeneye (*Bucephala clangula americana*) is slightly larger though almost identical in plumage.

The third member of the family, the Barrow's Goldeneye (*Bucephala islandica*), comes from Iceland, Greenland and North America, and though roughly resembling the other two at a distance there are several distinct differences in plumage when seen close to, besides being the largest of the group. The head is dark, sheened with deep purple, and the white eye-patch of the drake is roughly crescent-shaped instead of round. The underparts are similarly white but the black back, instead of being striped, has a single line of white 'windows' running along each side rather like the portholes of a ship. It is rather reluctant to breed in captivity and the price of a pair reflects this.

The most common member of the eider family kept in captivity, the European Eider Duck (*Somateria mollissima*) is possibly best known for the very soft down with which it copiously lines its nest, the original 'eiderdown' still collected and used for stuffing quilts, mattresses and insulating arctic clothing. Breeding in Scotland, Northern Ireland, Iceland and Scandinavia, the European eider is gradually increasing its breeding range southwards and in winter can be found all around the British coastline. It ventures inland only rarely, remaining out at sea for most of the time, where it lives mainly on a diet of marine molluscs and crustaceans, which it collects by diving. With its enormous feet set well back on the body it is uncomfortable ashore but perfectly adapted to water in which it dives deep, using its wings in a penguin-like fashion to propel itself deep under water in search of food. It therefore needs to be kept on deep water and given a richer, meatier diet than most other fowl.

The drake is a contrasting mixture of black and white, the upper body and breast creamy white with black belly, flanks and tail coverts and a noticeable white thigh-patch. The cheeks and neck are white, the crown black with a white stripe running along the centre, below which a smudge of light green runs down the nape of the neck. The bill is green, with a light blue-grey band near the tip culminating in a strong curved nail. At the base of the beak the green cere extends well up the forehead, merging into the typical wedge-shaped head of the species. The drake is quite vocal, particularly during his courtship display when he throws back his head and utters a series of loud cooing noises to attract attention. The female has a low grunt. Her plumage is a uniformly rich brown mottled with black. She hollows out a nesting scrape on a rocky shore or island, and lines it with down. They often nest colonially, and when incubating the four to six olive-green eggs, she only leaves the nest occasionally to feed throughout the 26 to 28 days incubation. Possibly because of this, when hatched many broods of youngsters are herded together to form large crêches, sometimes numbering hundreds of ducklings, which are attended by a few females only, leaving the others free to feed for long periods to build up the reserves of fat lost during incubation. By this time the drakes have long since departed for moulting.

The eider is a glutton in captivity and will gorge itself to excess if allowed, often becoming tame enough to accept food from the hand. They need a protein-rich diet but even then breed somewhat irregularly in captivity. The price of European eiders has risen steeply over the last few years.

Besides having four sub-species, the European eider has some very aristocratic relations, all of which are at present rare in captivity, needing good water, specialized feeding techniques, and a very deep pocket. The most ornate is the King Eider (*Somateria spectabilis*) which, although loosely resembling the European in body colour except for its black back, has a most colourful head with a crown and nape of light powder-blue contrasting sharply with a curious prominent orange bill-shield and bright-red bill. It lives on the very edge of the northern ice pack and nests in the arctic tundra.

Another very attractive sea duck from the north although also not for the beginner, is the Long-tailed Duck (*Clangula*

King Eider (*Somateria spectabilis*)

hyemalis), often referred to as the Old Squaw. Spending the summer in northern climes, it appears around the coastlines of northern Britain in winter. The birds are unusual in having two distinct plumages besides the eclipse. In winter dress the drake appears mainly white, with black breast, wings, face-patch and elongated central tail feathers giving it a piebald appearance. It has a pale grey eye-patch and the black bill has a broad pink tip with a black nail. The legs and feet are grey. In summer the black breast is retained, the head also turning black except for the grey and white face-patch. The white scapulars of winter are replaced by flowing chestnut feathers streaked with black. The long-tailed is at present found in only a few collections, its specialized feeding requirements and prohibitive price making it a bird for the specialist keeper.

Also diving ducks, but in a group of their own, are the sawbills or mergansers, a little tribe distinguished by their narrow, serrated bills with which they capture their main diet of live food: fish, crayfish and other small crustaceans, and also a wide variety of insects and larvae. With such varied dietary requirements the sawbills are not easy to maintain in good condition in captivity unless one has an area of water rich in natural food with which they can supplement their diet. They should certainly be avoided by the novice.

Possibly the easiest of the sawbill family to keep are the Hooded Mergansers (*Mergus cucullatus*), natives of North America. The males in breeding condition are most striking and attractive birds, mainly black above with rich cinnamon flanks, a white breast criss-crossed with black, and a beautiful erectile crest which gives the bird its name, a window of pure white bordered by black which is spread and closed like a fan. Its black face surrounds a vivid-yellow eye and the areas of black are sheened with green. The female is more quietly dressed, her browny-grey body, lighter on the underside, highlighted by a crest of orange brown.

The drake's display is a delight to watch. He inflates his neck with crest erect, shakes his head while rearing up on the water, and gives a vibrating throaty purr as he bows forward. The sound is somewhat frog-like and a group displaying in unison is a most endearing sight. The female responds by bobbing and jerking her head upwards and is soon lying prone on the water awaiting the drake's attentions. My own little flock breed quite freely, depositing eggs amongst other clutches, always in a nest-box, for in the wild they utilize hollow trees and old woodpecker nest-holes. That they breed so readily is almost certainly due to an abundance of life in the water, which I supplement by gathering bundles of underwater weed from a nearby marsh drain, which is full of aquatic invertebrates and larvae for which the birds will sift and dibble for hours. They are able to take a great many fish fry in the shallows and are a delight to watch winkling out tiny loach and minnows from between the stones on the gravelly bottom.

The almost round, white and shiny eggs are difficult to hatch, almost impossible using an incubator alone, their thick and extremely tough shells needing the assistance of some form of natural incubation. I often use mallard as fosters, for the hooded is also an early nester and there are always a few mallard on eggs at the time who fail to notice when the eggs are switched while they are off feeding. Having been incubated naturally, the eggs are then gathered up a day or so before they are due to hatch, placed in the hatcher and given really humid conditions, though even then they often require assistance in breaking out of the shells at 30 to 32 days of incubation. Normally hatched in a tree-hole, the young are natural escapologists as it is their first instinct to scale the inside of the hole to clamber from the nest. They are no less agile in a brooder, using even the slightest foothold to aid an astonishing climbing ability, and unless a brooder is covered they will often escape. The ducklings are by no means easy to rear, requiring plenty of live food in the form of high-protein mealworms. Grated hard-boiled egg is another requirement, sprinkled on a dish of chick crumbs. These have to be ground to a fine powder during the ducklings' first three weeks or so of life, as in its normal crumb form it apparently compacts in their gizzards and they soon wither and die. Surviving this period they will become very tame, their addiction to mealworms overcoming all fear; they will take them from the hand at feeding time. They quickly learn to come running as soon as the mealworm tin is rattled. All in all a rather difficult and unobliging species, but well worth the time and trouble of catering for them.

The largest of the mergansers, the Goosander (*Mergus merganser*) is the most attractive of all the sawbill ducks. The male is a quite spectacular bird, its black head and mane sheened with deep bottle green, the shoulders black and the creamy-white breast tinged with a delicate salmon pink. The rump and upper tail coverts are steely blue-grey. The slender, serrated bill, designed for grasping and holding fish, is bright red with a black culmen line running from the pronounced black nail to the forehead. The legs and feet are orangy red. A broad white wing-patch is prominent in flight, as are the black primaries. The female has steel-grey upperparts, flanks and tail coverts; her breast and belly are creamy white. Her head is shaggier than the drake, reddish brown with a white patch on the throat at the base of the lower mandible, which is similar in colour to the male's. Although quite large the birds have a distinctly streamlined appearance, built for diving and underwater mobility. The goosander lives almost entirely on fish, but will occasionally take shrimps, small crabs and mussels. Captive birds can be given whitebait and sand eels to enrich the high-protein diet necessary to keep them in good condition, and the goosander ideally needs to be kept on a large expanse of natural water where it can improve its diet with fish.

Similar in many respects to the goosander, though of a slimmer build, the male Red-breasted Merganser (*Mergus serrator*) can best be distinguished by the wide brown breaststripe studded with black, its black shoulder with white 'windows', and the ragged crest which gives it a decidedly unkempt appearance. Its flanks are vermiculated grey and it has a bold red eye. The females of the two species are less easily separated, though the merganser has a

Smew (*Mergus albellus*)

shaggier head with wispy crest. The red-breasted is unusual in that it uses both its feet and wings for underwater propulsion in pursuit of its main food – fish.

Widespread across the northern hemisphere, the southern limits of its breeding range includes the British Isles, though it is only occasionally to be found in collections.

Another delightful little sawbill is the Smew (*Mergus albellus*) a bird breeding in Scandinavia, Russia and Siberia which spreads westwards to Britain, mainly to the south-east, in small numbers during the winter. Highly desirable though scarce in collections, the male in full breeding dress is unmistakable, being mainly white with fine grey vermiculations along its flanks and a black mantle running to grey on the rump. The white erectile crest used in display partly conceals a black crown. It has a black facial mask around the eye and twin breast-stripes of the same contrasting colour. The bill and legs are dark blue grey. In eclipse he resembles the female and juvenile in the 'red-head' plumage. The female is almost grebe-like, with rusty-coloured head, white cheeks and throat and a dark patch around the eye in winter. Her back, flanks and rump are grey. In flight, both sexes display a white wing-bar. Fish form the greater part of the wild smew's diet, but a certain amount of molluscs, crustaceans and insect larvae are also taken, and thus in captivity the birds need a protein-rich diet.

Smew breed only occasionally in captivity, using a nest-box as a substitute for its normal nesting site in an old hole left by the black woodpecker, or in a tree crevice. Always in demand with collectors, its unique beauty and the scarcity created by breeding difficulties, make it expensive.

Another section of the diving ducks is the family of so-called stifftails. Two from the dozen or so species, the North American Ruddy Duck (*Oxyura jamaicensis*) and the Argentine Ruddy Duck (*Oxyura vittata*) are now becoming commoner in collections.

The drake North American is a brilliantly coloured bird, with a chestnut-red body, black forehead, crown and nape, white cheeks and a vivid sky-blue bill. The female is plain by comparison, her upperparts mainly a mottled mixture of dark grey-brown, apart from the belly, cheeks and upper neck, which are a mottled silvery white. She has a dark facial stripe running through her cheeks. Both birds are very small, short and stocky with tiny wings and typical stiff tail which is raised and lowered according to its moods. They walk with some difficulty, seldom leaving the water, where they dive frequently and well. The rear-positioned legs make the bird almost grebe-like in appearance.

The drake has a delightful display during which he erects his crest, which resembles a pair of short horns, one on either side of the head, cocks and fans his tail, inflates his neck and while expelling the air beats his broad bill rapidly against his chest, producing a drumming sound which can be heard from afar and causes a ring of tiny bubbles to rise from the water. A number of males displaying in unison is one of the most charming sights of the waterfowl world, as they work one another into a frenzy of activity while the females often drift around showing little apparent interest.

Unfortunately the eggs, up to nine or ten, are exceedingly difficult to hatch, most breeders preferring to leave them with their natural parent until incubation is almost complete, collecting them a day or so in advance of hatching or attempting to catch up the youngsters as day-olds for pinioning. For this reason a feral population has established itself in Britain, for if the ducklings hatch and make it to water they are extremely adept at diving and often resist all attempts to capture them. Thus a few full-winged birds have escaped from collections and in some areas feral colonies are being increased by wild-bred youngsters.

The ruddy duck has an eclipse plumage in winter, often not assuming full colour until late in the spring, though as they remain in breeding dress until late autumn they will add a splash of colour and charm to any collection while the rest of the northern-hemisphere fowl are in eclipse. Ruddy ducks are reasonably easy to keep on natural water though they are sometimes timid with other larger fowl. They will greatly benefit from a good quantity of water-weed among which they find a supply of aquatic life. When I find the time to dredge an armful or two of weed from the nearby marsh my ruddies in particular will dibble among it for hours.

The Argentine ruddy is far less common in captivity and is therefore more expensive than its northern cousin. With an almost identical body colour, it differs by having a sooty-black head and slightly narrower bill.

5

Wild Geese

Wᴉᴛʜ ᴀ ꜰᴇᴡ ᴇxᴄᴇᴘᴛɪᴏɴꜱ, the forty-odd species of wild geese found throughout the world can be classified into three distinct groups; the 'grey' and 'white' geese of the genus *Anser*, the so-called 'black' geese of the genus *Branta*, and the third group, the sheldgeese, that live mainly in the southern hemisphere and form part of the Tadornini tribe which also includes the shelducks.

Under the right conditions, geese can be very long-lived. Most species are easily kept, and in a fair-sized enclosure several species of northern-hemisphere birds can be penned together, for although they will stake out a territory and become noisier and apparently more aggressive during the breeding season, most of this show of aggression is bluff and actual conflict is rare. The same is certainly not true of the latter group, who have a far more belligerent nature and generally need to be kept well apart from other fowl to avoid hostilities.

Geese have few requirements. Provided one has a reasonable area of good grass and a small pool of water they will require little else apart from a daily handful of grain, and a few titbits of greenery during the winter months when grass becomes scarce. In return, apart from their aesthetic qualities, geese are extremely good watchdogs in an orchard or large garden, setting up noisily if an intruder enters their territory; and being primarily grazers one will be saved the time and effort of grass-cutting if a sufficient number is kept to manage an area of ground. With luck, certainly with the commoner species, most pairs will pay for their keep by settling down to nest and rearing an annual batch of goslings with the minimum of fuss once they reach adulthood at three years of age.

If for some reason goslings are hand-reared they will quickly become imprinted on their keeper. The first batch of Hawaiian goslings I reared in this way were collected as eggs during a heavy snowstorm a couple of days before they were due to hatch in the late spring. Not daring to leave them in such inclement weather, I transferred them to an incubator, and then carefully hand-reared them under a heating lamp. As I spent a great deal of time with the youngsters they soon became strongly imprinted on me, and when later released on a pond they would cackle a loud greeting whenever I appeared in sight and follow closely on my heels as I did the feeding round. If I sat down on the bank to watch the rest of the fowl my charges would follow, clambering all over me, nuzzling into my jacket, pulling my hair and eventually settling down comfortably in my lap, where they would often go to sleep. They seemed quite put out when I failed to follow them on to the pond!

Probably the most important member of the wild goose tribe is the Greylag (*Anser anser anser*), which has been kept in captivity for centuries. It is from this bird that almost all forms of domestic geese have evolved; in fact all but the domestic Chinese Goose which is a descendant of the wild Swan Goose (*Anser cygnoides*). The largest European grey goose, the greylag is not particularly common in ornamental collections, being rather cumbersome in build and not particularly attractive. Its body colour is a general greyish-brown, lightly barred and with noticeable white tail coverts. Its heavy bill is orange, while the legs and feet are a deep flesh pink.

As with all geese their main diet is grass, though as I found to my cost a party of greylags can be quite destructive to young trees or any other greenery to which they have access. Rearing a family of six in an orchard, I noticed a couple of newly planted apple trees were beginning to wither and die, and on closer inspection found the geese had completely stripped the trunks and branches of bark to a height of 3 ft (90 cm).

Greylags are now in common sight in my part of central Norfolk where in just a few seasons their numbers have increased from a few individuals to large flocks which flight at dawn and dusk to graze the pastures and cornfields. They have increased to such an extent that shooting has become necessary to keep damage to growing crops down to a bearable level. They provide a truly magnificent sight as they come cackling along the valley in wide V formations at first light, gliding down on arched pinions as the feeding ground is reached, where a few sentries keep constant vigil as the main army of birds take their fill.

There is an Asiatic form of the greylag, the Eastern Greylag (*Anser anser rubrirostris*) which differs from the western by its slightly lighter plumage and prominent pink bill. The greylag lays its five or six eggs in April or May, and after about 28 days of incubation the yellow-brown goslings will appear, taking about eight weeks to fledge fully.

Smaller and of a slimmer build than the greylag, the Pink-footed Goose (*Anser brachyrhyncus*) is a common winter visitor to Britain. As its name suggests it has bright pink legs and feet, and a bar of the same colour across its dark bill. The general body colour is grey brown, lighter on the underside and darker on the head. The flank feathers and back are edged with off-white and it has a grey wing-bar prominent in flight.

Pink-feet sometimes take time to settle down and breed in captivity, but when they eventually decide to raise a family the nest, a depression lined with grasses and down, is usually sited in an elevated position to allow a clear view of its surroundings. Laid in spring, the average clutch consists of five white eggs which hatch at 28 days, though occasionally six or seven are recorded.

Visiting Britain only in small numbers among flocks of migrating pink-feet, the Western Bean Goose (*Anser fabilis fabilis*) is of a similar overall colouring though slightly browner, but can best be singled out by its orange legs and feet and black bill with an orange instead of pink band. The extent of bill colouring is variable among the races, the western bean having several closely related sub-species which show geographical variations in this respect.

Bean geese are not widely kept, for although taming quite readily the bird is a reluctant breeder in confinement. The nest, a depression lined with vegetation raked together from local materials, is placed on a dry hummock or at the base of a tree. The five or six eggs are white and hatch after four weeks.

Another somewhat more attractive grey goose is the European White-fronted Goose (*Anser albifrons*), its pink bill highlighted by the noticeable white forehead which gives it its name. The grey-brown underparts are barred heavily with black across the lower belly. Legs and feet are orange.

Two races visit Britain's shores regularly in winter,

albifrons travelling south-west from its breeding grounds in Siberia, and the less common Greenland White-fronted Goose (*Anser albifrons flavirostris*), which flies south-east from its summer quarters in western Greenland, wintering mainly in Ireland and along the west coast of Scotland, though a few penetrate as far as the west coast of England and Wales. The Greenland race, also kept in captivity though in far smaller numbers, is readily distinguished by an overall darker appearance and light orange bill.

White-fronted geese tame readily and breed quite freely, often seeming to prefer an elevated nesting site, choosing a grassy ridge or bank on which to form the rough nest-scrape. The clutch of four to seven eggs is incubated for about 28 days and the juveniles when fully fledged are easily separated from the adults by the smaller patch of white on the forehead and lack of barring on the belly.

Far less common in the wild is the charming little Lesser White-fronted Goose (*Anser erythropus*), an even more desirable goose in collections for it is highly sociable and will live agreeably among any other fowl. It has a more petite and neater appearance than its larger cousins, and the darker plumage shows up the forehead-patch more prominently. Apart from the difference in size, closer inspection will reveal a bright-yellow circle of skin around the eye, a characteristic lacking in other members of the family. It is a very rare visitor to the British Isles in winter, only the odd straggler being found among large parties of European white-fronts. Although common in collections it breeds rather less easily.

The largest of the three white geese, the Greater Snow Goose (*Anser caerulescens atlanticus*) breeds in Greenland and its adjacent easterly islands, migrating to Canada and North America for the winter months. It is a mainly white bird with contrasting black primaries, the coverts of which are light grey. The legs and feet are fleshy pink, the heavy bill reddish pink. Snow geese stand out well in a mixed collection and all species are of an agreeable nature except when defending the nest site. The greater lays its six or seven eggs in a preferably elevated position and the clutch takes approximately 26 days to hatch.

Almost identical to the greater in its white form, the Lesser Snow Goose (*Anser caerulescens caerulescens*) has an additional colour phase, when it is sometimes referred to as the blue snow. Blue-phase birds vary, the first type being mainly blue-black with a white head and nape, the other form is similarly coloured but with an all-white breast. All colour phases have the typical reddish bills and flesh-coloured legs and feet. Pairing and interbreeding sometimes take place between the colour phases, possibly accounting for the somewhat variable plumage.

The smallest, and in many ways the most desirable, of the snow geese is the dainty Ross's Goose (*Anser rossi*), identical in plumage to the greater but with a blue base to the pink upper mandible. Birds of charming habits, the Ross's deserve a place in any collection for they tame readily and live agreeably with smaller ducks and geese even in a confined space, and provide a nice contrast in colour. Under the right conditions Ross's breed well in confinement, the nest consisting of a depression lined with grass, mosses and down, usually containing four or five eggs. The incubation period is 23 days.

Another attractively marked goose, the Bar-headed Goose (*Anser indicus*) breeds in central Asia and flies south to spend the winter in India. The general body colour is a pale blue-grey with white edges to the feather margins, a white head with two black bars extending to the cheeks, and a white stripe running down the almost black neck. The tail coverts

Ross's Snow Geese (*Anser rossi*) contrasting with Brent and Red-breasted geese

are white and the feet and legs orange, as is the bill, which has a black nail. Like most geese the sexes are similarily marked. The bar-head is quite ornamental though often remains rather shy and flighty in captivity, but even so breeds reasonably well. The incubation period is roughly 28 days. First-year juveniles are greyer and lack the white neck-stripe and black barring on the head, having instead white cheeks and a black crown and nape.

A popular bird among fanciers, the stockily built Emperor Goose (*Anser canagicus*) tames down easily and gets on well with other birds. The main body plumage is a bright blue-grey vermiculated with black and white in an attractive pattern, the amount of black greater down the front of the neck and under the throat. It has a white head and tail, the white extending down the back of the neck to the mantle.

The bill is pink with a black base and nail. Feet and legs are orange.

Restricted to eastern Siberia and Alaska, the emperor is quite rare in the wild but has become increasingly popular in collections. Breeding success is rather variable, the female laying her five or six creamy-white eggs in late May or June, and incubating them for about 26 days. Fully-fledged juveniles are much duller and lack the white head of the adult.

Although it is the ancestor of the quite commonly kept domestic Chinese Goose, the Swan Goose (*Anser cygnoides*) from Siberia is seen in captivity rather less than its descendant. The bird can be recognized by its very long and heavy black bill, which looks out of proportion to the rest of its body and makes it rather less appealing than many other geese. The top of the head is dark brown, the colour extending through the nape to the mantle, the cheeks and the forepart of the neck off-white. The general body colour is grey brown with lighter edging to the feathers on the back, while the breast is a warmer shade of brown. The feet and legs are orange. Eggs are white, five to eight, and take 28 days to hatch.

Of all the wild geese, the beginner could do little better than to start with a pair of Barnacle Geese (*Branta leucopsis*) being cheap, good-natured and easy to look after and breed, as well as being one of the most attractive of the over-wintering geese. As with the majority of northern geese the plumage of the sexes is identical, an attractive mixture of black, white and varying shades of grey. The white face has a black line running from beak to eye, the crown, neck and chest are glossy black and the belly and flanks light grey. The back is a beautiful slate-grey barred with black, the coverts edged with white. Legs, feet, bill and tail are black.

Barnacle Goose (*Branta leucopsis*) guarding female on nest

By their third spring a pair will be fully mature and will almost certainly attempt to breed. My own pair always chose the highest point in their enclosure for the nest, well away from other fowl and with a good, uninterrupted view of their surroundings, possibly following the instincts of their wild northern brethren who often pick a vantage point on inaccessible cliffs where they are safe from the attentions of prowling arctic foxes. The nest itself is little more than a shallow scrape lined with local vegetation and down in which the five or six white eggs are laid in late spring on alternate days. Directly the first egg is laid both birds guard the nest site avidly, and after about 25 days the fluffy blue-grey youngsters appear, possibly the most attractive of all young goslings. Like most geese, barnacles can be left to hatch and raise their youngsters and make excellent parents. The rearing pen should have a good supply of short grass for grazing and chick crumbs should be provided before weaning to pellets and wheat.

Two species of Brent geese are found in captivity, the Pale-bellied or Atlantic Brent (*Branta bernicula hrota*) and the Dark-bellied or Pacific Brent (*Branta bernicula nigricans*). Both species are small and neat and have black heads, necks, upperparts, feet and bills, a white neck-collar and white tail coverts, varying only in the depth of colouring on the breast and flanks and in the extent of the white collar. In the wild both birds frequent coastal regions, spending much of their time on the sea and feeding with the tide as it uncovers the beds of eel grass (*Zostera*), their staple diet. In captivity they require a good area of grazing and given the right conditions will become very tame, though for some reason they remain difficult to breed. This accounts for their high price. The eggs are usually laid during early summer in a depression copiously lined with down. They average five in number, are creamy white and take about 25 days to hatch.

Of about ten different species of Canada goose available, the most familiar is the Atlantic Canada (*Branta canadensis canadensis*), one of the larger types, which was first introduced to England in St James's Park in London during the seventeenth century and has now spread to become well established in many parts of Britain. A freshwater goose, it is a common bird of inland lakes and reservoirs, notably in East Anglia, where large numbers have to be culled, so widespread and prolific has it become that it causes considerable damage to agriculture. With flocks of up to 3000 recorded in some areas, the damage to growing crops can be severe.

Attaining maturity at three years of age, the Canada breeds freely on islands and beside water, and clutches of eight or nine eggs are not uncommon. One nest I discovered recently held no less than eleven eggs. A large goose, the Atlantic Canada requires plenty of space, spending much of its time on water and often only coming ashore to graze. The bird has also become well established in many town parks where there are suitable areas of water, quickly losing all fear of man and often becoming tame enough to feed from the hand.

The group ranges in size from the diminutive Cackling Canada (*Branta canadensis minima*), which weighs about the same as a mallard duck, to the Giant Canada (*Branta canadensis maxima*), which can weigh anything up to 14 lb (6.35 kg). The plumage of all types is basically the same, varying only in the degree of darkness of the main body colouring. All have black heads and necks, with a prominent white cheek-patch extending across the throat, black primaries and tail and a grey-brown back. The undersides vary from light buff-grey to dark sooty-brown. All Canadas are quite vocal and make good watchdogs, their notes varying from a husky cackling to a musical bell-like honking.

The most beautiful of all the wild geese, with striking coloration and an agreeable temperament, the Red-breasted Goose (*Branta ruficollis*) is the most sought after species for ornamental collections, and a little flock can be a great asset if a regular breeding programme can be achieved. The exceedingly attractive and distinctive plumage of black, white and rich chestnut-red makes a unique and exquisite bird that will become the centre-piece of any collection.

In the wild, the red-breasted breeds in the Siberian tundra regions, moving south in the autumn to the area of the Caspian Sea for the winter months, though an odd straggler can end up almost anywhere, even occasionally venturing far enough west to turn up in the British Isles among flocks of migrating white-fronted geese.

The red-breasted will quickly become tame and confiding, and although not easy to breed, once they are provided with the right conditions to get them started will usually produce a clutch each season. Their only drawback may then become evident, for the birds are quite vocal and will chatter incessantly if something disturbs them or threatens to enter the nesting territory. Curiously, in the wild the birds often chose to nest communally in the territory of a large bird of prey, usually a peregrine falcon or buzzard, presumably to receive some added protection as the raptors will mob and drive away predatory species of skuas and gulls and even arctic foxes when protecting the eyrie, thus also giving protection to the geese's eggs and young. My own red-breasted invariably chose to place their nests as close as possible to my snowy owl pen, some latent instinct possibly prompting such behaviour, though there are many more likely-looking nest sites in their enclosure. The eggs, a light olive-brown, and from three to seven in a clutch, take about 25 days to hatch. I collect each egg as it is laid, substituting it with a dummy, for the shells are extremely thin and will not withstand much shuffling around in the nest before they crack. The geese add to the danger by their annoying habit of adding stones to the nest. With such expensive eggs it is a worthwhile precaution. The goslings hatch well in an incubator and are dark brown with greenish-yellow underparts, but within a few weeks the distinctive red breast feathers will appear through the down and they are fully fledged in about eight weeks. Being a much sought-after goose, red-breasted remain rather expensive to buy.

The Hawaiian Goose or Ne-ne (*Branta sandvicensis*) is one

of the success stories of captive breeding projects, for as recently as the early 1950s it was on the very brink of extinction until a few birds were acquired for breeding, notably by the Wildfowl Trust in Gloucestershire, England, from the forty or so individuals remaining in the wild. Fortunately the ne-ne was found to breed quite readily in captivity, and following a few years building up a healthy captive stock, many were taken back and reintroduced to their native islands, which now support a generous and thriving population. Many more are kept in collections throughout the world, and the future of this goose now seems assured by the timely rescue operation.

The ne-ne is an attractive goose, with cheeks and neck of creamy-yellow buff edged with a face and crown of black which extends down the back of the neck. Body colour is grey brown, barred and streaked with patches of light and dark. A most unusual feature is the significantly reduced area of webbing between the toes, an evolutionary characteristic which suits its rather more terrestrial life on the volcanic uplands of its native islands.

In captivity the gander can become quite aggressive, especially at breeding time when it may bully other birds that venture too close to its mate or to the vicinity of the nesting site. It is therefore possibly best kept on its own, or in a large enclosure where other fowl have the opportunity of giving it a wide berth. My own gander has a peculiar habit of beating up his female if one approaches too closely, being if anything rather over-protective towards her, though this is possibly all to the good. Left to their own devices a pair will usually rear their own young quite successfully. They are early nesters, the clutch often being complete in late winter with endearing grey goslings appearing some 30 days later.

Ne-nes are a worthwhile addition to any collection, for besides the pleasure of keeping them one can get a certain satisfaction when they eventually breed from making a contribution, however small, towards their future security as a species.

While containing some of the most attractively coloured geese, the southern-hemisphere group known as the sheldgeese are related to the shelducks and as such are unsuitable for housing in a mixed collection unless in a large enclosure. They are safer penned separately, for their beauty is only skin deep and belies their aggressive nature. Like shelduck, most species are quarrelsome and pugnacious, particularly during the breeding season when they become very bad-tempered, even to the extent of causing injuries to smaller fowl.

One of the most colourful is the Egyptian Goose (*Alopochen aegyptiacus*), an extremely handsome though aggressive goose that really looks the part, its dark-chestnut eye-patch giving it a countenance that betrays its true disposition. The main body colour is orange buff with a chestnut breastplate and neck markings, richer brown above, with black primaries, metallic-green secondaries and bold white wing coverts most noticeable in flight. It is supported in a rather upright posture by bright-pink legs. A native of Africa, in Britain its increasing feral population is confined to East Anglia, and in my particular part of Norfolk there are many that live on and around the large lakes formed recently by gravel extraction along the river valley, from where they flight out to the nearby water-meadows and fields of young corn to graze. I have recorded flocks of up to 60 individuals feeding on the marsh adjoining my duckponds, though they normally prefer to move around in groups of five to ten, or in well-defined pairs that are always fighting amongst themselves and disputing territory with rival pairs as the time for breeding comes.

When nesting, Egyptian geese will often utilize natural tree crevices or a platform in a fork of spreading branches for their nest. The birds have an extended breeding season, for feral youngsters can appear from late winter to mid-summer. The goslings closely resemble shelducklings, pied dark brown and white. I have watched broods parachuting down to earth from high trees soon after hatching as the parents wait below calling anxiously. There can be as many as nine in a clutch, though by the time they are reared the number is often significantly reduced as they are taken by feeding pike during their first few vulnerable days of life on the nearby lakes.

A small-billed goose, the Ashy-headed Goose (*Chloephaga poliocephala*) comes from southern South America. Apart from the larger size of the male the sexes are alike, the head and neck ash grey, the breast bright rufous and the white belly extending to the flanks, which are finely barred with black. The upperparts are grey brown with light rufous and black barring on the scapulars. The forewing and secondaries are white, separated by a black speculum sheened with metallic green. Primaries and tail are black. The legs and feet are unusual in being bright orange with black mottling on the front of the tarsus, toes and webs. Although quite small, they can become bad-tempered towards other fowl during the breeding season, when the nest is built among light vegetation. Incubation of the four to six light brown eggs takes about 30 days.

Also from the southern tip of South America, the Ruddy-headed Goose (*Chloephaga rubidiceps*) is becoming rather scarce in the wild, though a reasonable population exists on the Falkland Islands. Noticeably smaller though of a similar built to the ashy-headed, both sexes have reddish-brown heads and breast and grey-brown upperparts; most of the body is finely barred with black. The wing is identical to the former, as are the legs and feet, though with rather less black mottling.

Noisy and aggressive, the ruddy-head is an irregular breeder in captivity but when successful it sites its nest of cream-coloured eggs among thick vegetation. The incubation period is around 30 days.

The largest of the sheldgeese, with a proportionately tiny bill and short, thick neck, the Andean Goose (*Chloephaga melanoptera*) is a species definitely to be kept in isolated pairs, being particularly aggressive in defence of its territory and positively dangerous to other fowl when protecting eggs and young. It is a striking bird, basically white with black primaries, tertiaries and tail. These dark areas are sheened lightly with green, purple and bronze, and the white scapular feathers are spotted with dark brown. The reddish legs are set well back to give it an upright stance, and the bill is pink with a black nail.

The male's display consists of a depressing of head and neck into the mantle, a ruffling of feathers and a high-pitched whistling which the female answers with a low cackling growl. Andeans need a good area of grazing to themselves but will be content with the mimimum of water; even the smallest of ponds is quite ample for their needs.

Distantly related to the Andean goose, in the wild the range of the Abyssinian Blue-winged Goose (*Cyanochen cyanopterus*) is restricted to the highlands of Ethiopia, though it is found in quite a few ornamental collections. Apart from the considerably greater size of the male the sexes are identical; basically grey brown, lighter on the head and neck and darkening on the breast where the feathers have pale centres, which gives the bird a marbled appearance. The upperparts are dark grey-brown, browner on the mantle. The relatively small bill is black, as are the rather short legs. The blue forewing is partly concealed when at rest but conspicious in flight, as are the white underwing coverts. The secondary feathers of adults are glossed with green, the primaries and tail black. Both sexes have a high-pitched whistle and a nasal barking note when alarmed.

The blue-wing is a small, stockily built goose that tames readily but sometimes takes to bullying birds smaller than itself. It can, however, be kept safely among larger species. It is largely nocturnal, becoming active at night and spending much of its time during the daylight hours resting, when it has a curious habit of supporting its head and neck on its mantle, even when walking.

The birds have an extended breeding season which in the wild runs through spring, but nests have also been found in summer, and even early winter, in its native habitat. In captivity the nest is often hidden away among quite dense vegetation or under a bush, where incubation of the four to six cream-coloured eggs takes about 30 days. The young are quite attractive, with yellow heads and necks, black upperparts with silvery markings and a black area behind the eye.

Found on the coastal grasslands and islands of southern Australia, the Cape Barren Goose (*Cereopsis novaehollandiae*) is another attractive bird present in only a few collections due to its anti-social behaviour. A heavily built goose, the basic colour is light blue-grey with dark spotting on the back and black edging to its flight feathers. It has a white forehead and crown, a reddish-brown eye and an enlarged lime-green cere extending almost to the tip of the black bill. The feet and legs are pinky red with blackish mottling. The birds have an almost pig-like grunt when annoyed or alarmed.

One of the most pugnacious of geese, the species must be penned separately to avoid conflict. It is particularly belligerent during the breeding season, and will even see off a human intruder from its territory with thrashing wings and spiteful beak.

6

The
Swans

ONE SHOULD CERTAINLY think twice about keeping any species of swan, for unless one has a great deal of space available few are entirely suitable for housing with a mixed collection of fowl. With the exception of the coscoroba from South America, swans are very large, heavily built and potentially spiteful and aggressive, and can create havoc among smaller species of waterfowl, particularly during the breeding season when they become highly parochial and will lay claim to a very large domain which is defended viciously against all comers. Most species also need to be provided with a large expanse of natural water where there is a constant supply of natural aquatic vegetation, the mainstay of their diet, which they dredge up from the bottom by beak-dipping and up-ending in shallow waters. Swans also need good grazing, and will welcome grain and greenstuff in the form of lettuces and lawn trimmings, and even household scraps and bread, the latter thrown on the water to soften.

Unless a large enclosure is available, no more than a single pair of swans should be kept together, for although they are somewhat gregarious during the winter months, come the spring each pair will occupy a large territory as the time for nesting approaches.

In all species male and female – 'cob' and 'pen' – are virtually identical in plumage, although the male is slightly heavier and of a stouter build. There is no seasonal variation in plumage. Swans are very long-lived, pair for life, and most species share in the task of nest-building and the rearing and guarding of youngsters. As breeding time approaches, a plentiful supply of suitable nesting material must be provided, for the nest consists of a huge mound of vegetation raked together, often many feet in diameter, which is often added to during incubation.

When nesting the birds are best given a wide berth and left mainly to their own devices, for an upset swan is a mean adversary and is capable of inflicting damage as it thrashes out with powerful wings at anything intruding on its province. Like geese, swans are better left to hatch and rear their own young and will usually perform the task admirably. The cygnets of some species, when first hatched, are often carried on their parents' backs while on the water to protect them from underwater predators.

Everyone is familiar with the Mute Swan (*Cygnus olor*) for it is probably the best known of all waterfowl, a symbol of grace and beauty with a tame and confiding nature. Semi-domesticated, it is easily approached, a bird of public parks, lakes, village ponds, streams, rivers and broads, where it spends much of its time in regular contact with man. Not really mute, it has curious croaking and trumpeting notes and an aggressive hiss when angry, when it will show its displeasure by arching its wings threateningly and appear to double in size. The largest British bird, the mute swan is sturdily built, often weighing on the heavy side of 30 lb (13.6 kg) and having an extended wingspan of over 7 ft (2.1 m).

The juvenile swan generally attains its completely white adult plumage following the second moult, when the bright orange bill is tipped with a black nail and at its base is a large black knob, more prominent in the cob. The feet and legs are black. Its powerful wings make a distinctive thrumming sound when in flight, audible for at least half a mile.

Adults are not fully mature for breeding until at least three years of age. They prefer an island nesting site, though will build, almost anywhere beside water, a huge nest of rushes, water-weed, grasses and bankside vegetation dragged together to form a mound that can measure up to 6 ft (1.8 m) across, in which the large eggs, greeny white at first but becoming stained with vegetation, are laid in mid-spring. These usually number six or seven, but I have found ten eggs in a nest and up to twelve have been recorded. The nesting site is often re-used each year by the same pair and is guarded jealousy by both birds as soon as they take up residence in the early spring.

The cygnets hatch following about five weeks of incubation, at first clothed in a buff-white down and guarded avidly until they are old enough to fend for themselves. Fully-fledged juveniles are grey-brown with pinky-grey bill and grey legs, often remaining under parental supervision until well into the following winter, when the birds become more gregarious and often join together to form small flocks at good feeding sites until the spring dispersal to breeding areas.

There are two other wild swans found in Britain, though both are winter visitors only. The Whooper Swan (*Cygnus cygnus*) gets its name from its bugle-like call, a musical note that carries afar. It is very vocal in flight. Slightly larger than the mute, it is an all-white bird with heavy yellow and black bill, the amount of yellow greater than the other British swan, the Bewick's, and extending along the sides towards the tip of the bill. The feet are black. The whooper also has a longer body and neck, and the longer bill gives the head a more pronounced angular shape.

If one has the space whoopers will do well in confinement and can breed quite freely provided plenty of nesting material is made available. The nest, a bulky accumulation of waterside and aquatic vegetation and grasses, is preferably sited on an island or close to the water and usually contains five or six creamy-white eggs that are laid in early summer. The young hatch after about five weeks, at first covered with a thick layer of white down with flesh-coloured legs and feet. The bill, also flesh-coloured, is darker towards the tip. The cygnets take about nine weeks to fledge fully.

Almost identical to the whooper, the Bewick's Swan (*Cygnus columbianus bewickii*) is less heavily built, has a shorter body and neck and infinitely variable beak markings.

After rearing their young, the birds leave their breeding grounds in Siberia to winter across Europe, the first small flocks turning up in Britain from October onwards, when their arrival is announced with wild bugling cries as they pass high across the sky in vee formations, one of the first signs of an approaching winter. Many spend the colder months eking out a living on estuaries and along the coastline, and a fair proportion end their journey at the headquarters of the Wildfowl Trust at Slimbridge in Gloucestershire, where, after discovering that each bird has its own particular beak pattern, details of several thousand individuals are kept, allowing records of their movements to be made. The extent and pattern of the yellow bill markings was found to be individually unique and almost every bird in the large flock that spends the winter at the refuge can be logged, a system comparable to fingerprinting.

The Bewick's is uncommon in captivity, but although needing less space than the more aggressive mute it is a reluctant breeder unless conditions are to its liking. There is an eastern form of the Bewick's called the Jankowski's Swan (*Cygnus columbianus jankowski*) but as the two races are said to interbreed there is some controversy as to whether or not they can be individually recognized.

Originating in Australia, the Black Swan (*Cygnus atratus*) has been successfully introduced to other countries, notably New Zealand, where there is now a well-established wild population that has at times to be controlled by shooting and egg collecting. Rather ungainly on land, the black swan is graceful and serene on water and is a familiar sight in zoological gardens and among larger collections of waterfowl. Its sooty-black plumage is edged with grey, though at a distance it appears completely black. The primary feathers and the outer half of the secondaries are white, hidden while at rest, and on its back the folded wing coverts and large tertiary feathers are curiously crinkled, giving the bird a rather frilly appearance. The strong bill is a waxy crimson with a broad white tip, the base skin extending to the reddish eye. The legs and feet are black. The bird has a high-pitched bugling call, quite musical when heard from a distance.

Highly gregarious in the wild, the black swan often nests colonially, but as it is inclined towards aggression is better kept apart from smaller fowl, though it needs rather less room than most other swans. When mature they breed quite freely, possibly producing two clutches during the extended breeding season when both male and female take turns to incubate the clutch of five to eight pale-green eggs, which take about 35 days to hatch. The cygnets are at first greyish brown with a grey-tipped black bill and occasionally hitch a

Black-necked Swans (*Cygnus melanocoryphus*)

ride between the arched wings of their parents while afloat.

A highly ornamental species, the Black-necked Swan (*Cygnus melanocoryphus*) is far less of a bully than other swans and in a fair-sized enclosure can usually be trusted among other birds with the exception of the shelduck, with which it will almost certainly pick a fight. The slight resemblance between the two birds possibly presents the shelduck as a potential rival and seems to be the cause of frequent disharmony between the species.

The plumage is unmistakable, the body white with velvety black head and neck, with a narrow white eye-stripe running upwards to the crown. The blue bill has a well-developed basal knob of bright red. The legs and feet are flesh pink.

When excited the bird has a wheezy though musical whistle, repeated frequently during the display as the neck is pumped up and down. Black-necked swans breed quite readily in captivity once mature, laying four to seven creamy-white eggs which may be replaced with a second clutch if the first is collected. They take about five weeks to hatch. The downy cygnets are white, their bills black with white bases and black legs and feet. For much of their early life they spend a great deal of their time riding on the backs of their swimming parents, a habit that sometimes remains until they are several weeks old.

The smallest of the family, the Coscoroba Swan (*Coscoroba coscoroba*) gets its name from its call, being rather un-swanlike in appearance, for it lacks the slender and graceful lines of the true swans and perhaps shows more affinity to the geese. Like geese, it prefers to graze rather than dredge up aquatic weed. There are suggestions that the bird is not too distantly removed from the Dendrocygnini tribe, the whistling ducks, which it resembles in its rather upright stance and duck-like bill. Both sexes are pure white apart from black tips to the outer primary feathers. The legs and feet are pink and the bill a bright waxy red with a lighter tip.

In the wild the coscoroba occupies much the same restricted area as the black-necked swan, a resident of southern South America, though only a visitor to the Falkland Islands. Lacking the aggressive nature of other swans it can safely be kept among a mixed collection without fear of trouble, though for some reason it remains a species still found in only a few collections. Breeding success is somewhat sporadic, possibly accounting in part for its scarcity, but when successful the eggs are incubated for about 35 days.

The two remaining species of swan are from North America, and neither are very widespread in captivity. The largest and rarest, with a wild population of only a few thousand, is the Trumpeter Swan (*Cygnus buccinator*) the largest of all the waterfowl and recognized apart from its massive size by the large, entirely black bill, which gives the head a wedge-shaped appearance. An all-white bird with massive black feet, the trumpeter needs a large area of water and is best kept in isolated pairs, for they can become nasty-tempered in the breeding season when they are liable to inflict damage on lesser fowl if penned together. They are, however, more kindly disposed to their keeper, and can become tame enough to feed from the hand.

The trumpeter has an elaborate and beautiful mating display, a ritual of head-dipping, bowing and trumpeting, culminating in a bout of wing-flapping and calling as mating is achieved on the water. The normal clutch is four or five white eggs, laid in a huge nest, which soon become stained with vegetation.

The smaller Whistling Swan (*Cygnus columbianus columbianus*) is the American counterpart of the Bewick's. It has a small but variable area of yellow at the base of the bill just forward of the eye. It also has black legs, the rest of the plumage being pure white. As in many of the white swans, the feathers of the head and neck often become stained rusty brown through frequent immersion in water as its dibbles for its main diet of aquatic underwater vegetation.

7

Breeding

IT SHOULD BE the ultimate aim of every waterfowl-keeper to encourage as many as possible of his stock to breed, for apart from confirming the fact that one has achieved a happy and healthy environment in which the birds are contented enough to settle down and attempt to raise a family, there are the added pleasures and rewards of producing a yearly batch of surplus fowl. These can be sold or exchanged for new varieties, thus recovering the initial outlay and making the hobby entirely self-supporting, and possibly even allowing a little profit to be made, though the time, labour and everyday running expenses will ensure that the amateur breeder is hardly ever likely to make a fortune.

A good diet and suitable environment are both essential for getting the best out of your waterfowl, and much can be done in the run-up to the breeding season to ensure the birds have every chance of producing a few clutches of fertile eggs. The better the breeding stock are fed and catered for, the more eggs they will produce with a subsequently higher rate of fertility.

A few weeks before the first eggs are expected, the normal

maintenance ration, possibly a fifty-fifty mixture of poultry-growers' pellets and wheat, can be gradually altered until one is feeding a diet consisting almost entirely of breeders' or layers' pellets, supplemented with high-protein additives such as biscuit meal containing meat, which will go a long way towards bringing the birds into a good breeding condition. The protein intake can also be boosted during the breeding season by feeding a few handfuls of trout-rearing pellets, the large floating type which can be thrown on the water. These consist of almost 50 per cent protein and the birds love them, but being rather expensive they are given in moderation. With trout pellets and other titbits in the form of aquatic weed dredged from the nearby marsh ditches, which harbours a surprising amount of animal life, I have been able to obtain a good supply of fertile eggs from most of my fowl, including moderate success with canvasbacks, scaup, goldeneyes and hooded mergansers, species not easily bred. Admittedly, their diet is enhanced by a supply of natural food always present in the ponds, brought in from the nearby marshes via the inlet channel flow.

If waterfowl are to be encouraged to breed successfully in confinement a varied selection of good nesting sites is essential, whenever possible as near to the natural situation favoured by the birds in the wild as can be achieved, and ideally there should be a choice of at least two sites for every pair in the collection. Many will simply make use of the available natural cover, and suitable areas should be left to grow unchecked to allow birds to select a site. If one has a large area of ground, rough areas can be narrowed down by judicious trimming to make the search for nests less of a chore, but blocks of vegetation must be left of a suitable size to make the search difficult for egg-eating predators. Most of the rubbish can be cleared in the autumn as soon as the laying season is over, though some must be left to provide nesting accommodation for the following spring.

Basically, three types of nesting sites are required for a mixed collection. Besides natural vegetation, a variety of nest-boxes and baskets are ideally suited to the tree hole nesters, and a selection of tunnels in the ground for species such as shelduck, which in the wild often take over disused rabbit burrows for their nests. Geese are the exception, and will usually select a spot completely in the open on a high area of ground or against a tree trunk, often in quite a prominent position. As they guard the nest constantly as soon as the first egg is laid, there is little need to conceal the nest from predators.

If natural ground cover is sparse, particularly when establishing a collection, a pile of fir boughs or hedge trimmings loosely heaped together can provide sufficient temporary accommodation for a nest, but in time much can be done to improve the breeding habitat by planting up an enclosure with a variety of ground cover, evergreens and shrubs, with an eye to both its aesthetic and practical use, many species of ornamental shrub and evergreen being eminently suitable for both purposes. Low ground cover is most important, and laurel, broom, box, horizontal conifers, yew, gorse and even holly will provide good nesting cover, besides enhancing the surroundings, and most can be shaped or pegged down while growing to form good natural sites. Lawson's cypresses are ideal if the tops are taken out before they grow too tall, to encourage growth at a lower level. The bottom branches trail to the ground, providing a dense screen around the trunk where the birds will be able to nest. Reeds and rushes can be planted in dense blocks beside the water; and in drier locations clumps of pampas grass are ideal, affording excellent cover once established besides being

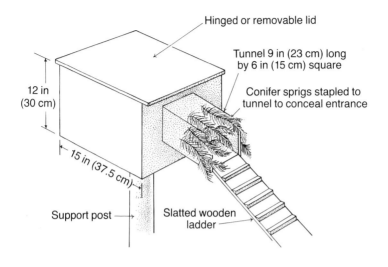

Hinged or removable lid

Tunnel 9 in (23 cm) long
by 6 in (15 cm) square

Conifer sprigs stapled to
tunnel to conceal entrance

12 in
(30 cm)

15 in (37.5 cm)

Support post

Slatted wooden
ladder

Fig. 7   Nest-box

highly ornamental. I have several wild clumps of dog's mercury growing at the waterside, an indigenous plant that is thick, grows to an ideal height and is much favoured for well-concealed nesting sites for many of the ground-nesting species, and I also leave useful clumps of nettles, wild teazle and bramble patches dotted about the enclosures.

Many birds will take readily to nest-boxes, particularly the tree-nesting species which in the wild favour natural tree crevices or old woodpecker holes large enough to accommodate the nest. Each box should have a short entrance tunnel to allow privacy and to discourage winged vermin from entering. I find that 15 in (37.5 cm) square × 1 ft (30 cm) high is an ideal size. The roof should slope slightly towards the back with a lift-off or hinged lid for easy inspection (Fig. 7). An entrance tunnel about 6 in (15 cm) square and 9 in (23 cm) long can be attached to the front, the bottom of the tunnel raised a few inches above the actual nest level to prevent any eggs from rolling out or being ejected by other birds wishing to use the nest. Almost any old wooden box can be converted to good use, but shuttering plywood is quite cheap and readily available from any builders' merchant. Solidly constructed and treated with preservative regularly, such boxes will last for many seasons. The boxes can be placed directly on the ground in suitable spots or raised on legs with a log or slatted ladder fixed against the entrance tunnel to allow access. Raised nest-boxes are particularly favoured by mandarins and carolinas when erected against the trunk of a tree to re-create their natural nesting site. It is better to provide as many choices of situation as possible.

A large proportion of my nest-boxes are fixed on platforms or legs directly over the ponds with ladders leading straight from the water, which prevents rats from finding or entering the nest. I find these are the most successful, for birds tend to feel more secure when they can enter a nest-box directly from the water. The box should be lined with a few inches of fine soil from which all the stones have been removed, or peat, and a handful of straw or dry grass added from which the duck will line her nesting scrape. The entrance tunnel can be garnished with a few sprigs of evergreen stapled around the entrance hole to increase privacy and help conceal the nest from egg-eating predators.

Interwoven nesting baskets (Fig. 8) specially designed for the purpose are successful though rather expensive, and are particularly useful if sited amongst natural cover, where they offer an extra measure of protection. They can also be raised on legs above the pond, and are very effective if placed just above water level but low enough for the birds to climb in

Nest-boxes built on a platform over the water

directly from the water. Their tunnel-like shape has the advantage of preventing all but the most determined of winged vermin from reaching the eggs.

Underground nesting chambers can be made easily using shuttering ply or rough boarding to make up a bottomless tunnel, the entrance 6 in (15 cm) square and the sides angled out to give a good-sized nesting chamber, it should be at least 15 in (37.5 cm) wide to allow plenty of room for the bird to turn around inside. Dug into a sloping bank, the whole tunnel is completely covered with soil apart from the entrance hole. The actual tunnel should be no more than 3 ft (90 cm) in depth, sufficient to conceal the nest but short enough to enable its contents to be removed easily. Here again all stones must be removed from the tunnel floor to prevent damage to the eggs, for even a tiny hair crack is sufficient to prevent an egg from hatching. However careful

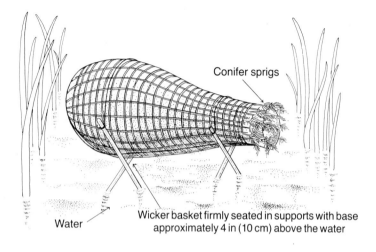

Conifer sprigs

Water

Wicker basket firmly seated in supports with base
approximately 4 in (10 cm) above the water

Fig. 8   Wicker nesting basket

one is, stones will often mysteriously appear. Ducks in particular have the annoying habit of adding them to their nests, and whenever possible these should be removed to avoid breakages. Add a good handful of straw to the nest and again camouflage the tunnel entrance with a few sprigs of evergreen. The duck will soon investigate and find the hole to her liking, but any crows or magpies intent on a feast of eggs will be discouraged as they are extremely reluctant to enter such confined spaces. So successful have these tunnels been that I have wild shelduck and mallard making regular use of them. An old length of large-diameter drainpipe dug into a sloping bank works just as well.

With a little imagination almost anything can be put to good use for nesting. Hollow logs, old tree stumps, even cleaned-out 5 gal (23 litre) oil drums with an entrance hole cut in one end will suffice, but on occasion even the most palatial of residences will sometimes be ignored while others, however undesirable-looking, will be re-used year after year. Most garden centres sell old wooden barrels cut in half for use as flower-tubs and these, upturned with an entrance hole cut in one side and a tunnel of twigs forming a concealed entrance, will make a good safe nest. Another cheap alternative is to construct a rough wigwam of sticks and boughs stuck into the ground in a circle large enough to hold a nest inside, or a section of old wire netting shaped in a rough dome, double-skinned and having a sandwich of vegetation and grasses between the two layers. Pegged down at the side and with the wire bent to form a small igloo-shaped dome, many such sites can be made for little or no outlay.

Spring is the most exciting time of the year for the waterfowl-keeper, with every bird decked out in its full breeding attire and the air full of the drake's attractive mating calls and beautiful courtship displays. Pairing up, displaying and even the act of copulation can begin as early as late autumn when the drakes have fully coloured, but in the spring just prior to laying the first egg, the duck will shuffle a rough scrape in the earth or inside a box with her breast, forming a dish-like depression which will later be lined with vegetation as the eggs are being laid. Often by searching slowly and carefully one may discover where a bird, particularly those of the ground-nesting species, has made several such scrapes within a few feet of one another before she finally decides which one to use, and deposits the first egg in the nest of her choice. By this time, a nest can occasionally be found by watching the birds rather sheepishly standing guard in the area. It is a good idea to acquire some dummy eggs to swap for the first ones laid – hens' eggs will do, having a more natural feel than the artificial egg – for during the

Clutch of eggs in a nest-box

first few days the females are quite careless with their eggs, often leaving them uncovered, offering an open invitation to predators. Once a few have been laid, the duck will begin to cover them carefully with vegetation as soon as she leaves the nest; and by the time the clutch has been almost completed she will line the nest with a blanket of soft down pulled from her breast, which helps the eggs retain their heat while she is off the nest for her daily feed. Once covered, the nest can be difficult to find. Often the only sign is a few pieces of wilting vegetation with which she has covered the eggs. If one has a lot of ground to cover it can be difficult to find every nest. The only answer is to make a systematic search, although, as ducks often lay in the early morning, while the dew is still on the grass one can follow the little secret trails through the foliage to discover where a bird is

Storing eggs in a tray lined with damp sand

laying. Often the path takes many little twists and turns along the way, so intent is the duck on concealing her nest.

If eggs are collected daily once laying has begun, they will need to be suitably stored until a clutch has been completed before setting for incubation. Always leave at least a couple of eggs, or dummies, in the nest to encourage the duck to keep laying. Luckily, ducks cannot count and will continue to add to the clutch until it is ready to incubate. When picking up, always remember that fertile eggs are extremely delicate objects and as such will not stand too much rough handling. A bucket lined with soft grass will stop them rattling around and being damaged during collection.

Correct storage of the eggs is all important. A good method of storing prior to incubation is to use a shallow tray lined with damp sand positioned in a cool room, which will prevent the eggs from drying out. Should they dry out the membrane enveloping the duckling inside the shell will become tough and leathery, and subsequently difficult for the duckling to puncture as it begins its struggle from the shell. It must also be remembered that all eggs should be turned at least twice daily through 90 degrees to keep the contents from settling.

Ducks' eggs should be set as soon as the clutch has been completed, for fertility begins to drop the longer they are stored. In situations where incubation has been allowed to continue with the parent bird, when collecting the 'hot' eggs they must be transferred to the incubator as quickly as possible without being allowed to cool. To retain their heat, cover them in a bucket with the wad of duck down that will be found in the nest. In emergencies I often collect odd clutches inside my shirt, my body heat being sufficient to keep them warm. However, this is not always a good idea, particularly as an incubating duck when disturbed will often defaecate over the entire clutch when leaving the nest, with rather unsavoury consequences. This act evidently masks the scent of the nest when a duck is disturbed by a predator, thereby possibly saving the clutch. I can certainly vouch for its effectiveness.

By providing a good diet, a peaceful and vermin-free environment and plenty of favourable nesting sites, one has done all it is possible to do to ensure successful breeding. The rest is up to the birds.

8

Hatching

Having induced your waterfowl to breed and produce a plentiful – and hopefully fertile – supply of eggs, you are then faced with the task of hatching them. Incubation is a complex subject, and 'Don't count your chickens . . .' applies equally to waterfowl, for consistently good hatching results can be very difficult to achieve and maintain.

The 'hatchability' of waterfowl eggs can vary enormously from bird to bird, even among the same species. A classic example of this is my two pairs of northern pintail, which are housed in the same enclosure under identical conditions and supplied with the same rations. In a good season both produce two clutches of eggs. The ducklings from one bird, once hatching begins, will all invariably start chipping in unison and emerge safely from their shells within an hour or so of one another. The eggs produced by the second bird always seem to cause problems, the hatch sometimes being prolonged over a full 24 hours after the first duckling has emerged, with some youngsters almost certainly requiring assistance to break free of the shells. What causes this variation has always remained something of a mystery.

It is seldom advisable to allow the parent duck to attempt to hatch and rear its own young, for although under her care a higher percentage of youngsters will often hatch out by the old tried and tested method, captive ornamental ducks rarely make good parents, particularly when many fowl are kept together in a fairly confined space. The parent will lead the ducklings constantly round and round the enclosure trying to escape from other fowl until they become thoroughly exhausted, and in the end she possibly loses the lot. Even in the wild, a large proportion of broods are often lost, although this is usually due to a combination of predation, lack of food and bad weather. Despite possibly hatching out the entire clutch of eggs in the wild, in many cases a duck will end up with only two or three fully-fledged birds to show for her efforts. Even nature needs an occasional helping hand. Coupled to this is the fact that many ducks will normally lay a replacement clutch or two if early eggs are picked up quickly, and in this way many more eggs can be produced per pair of birds, which does much to improve the balance sheet at the end of the season. By collecting and incubating the eggs artificially, my pair of marbled teal produced a total of 44 eggs one season from which well over 30 young teal were reared, a very good output from a single pair of birds. Geese and swans are an exception to the rule, at least with any second clutches, and if left to their own devices will usually hatch and rear their own youngsters quite competently, protecting them directly the first egg is laid and seeing off any form of danger from the nesting site or the young goslings. But such is certainly not the case with ducks, who need an alternative method of propagation.

Not so many years ago the good old broody hen was used almost exclusively for game- and waterfowl-rearing, achieving high hatchability and later teaching her charges the ways of the world, showing them how to feed, protecting them from danger and providing warmth and shelter from the cold, in fact catering for all their needs. It has now become increasingly difficult to obtain good breeds of hen for this purpose unless one is prepared to keep them oneself, for the majority of modern fowl are bred to be little more than animated egg-machines, and in the selective breeding process most have lost much of their maternal instinct, being highly strung, temperamental and often quite unpredictable. The slightest disturbance can upset them; even a sudden change of weather is enough to unsettle an apparently tightly sitting hen. I lost several clutches of eggs one spring afternoon following a severe thunderstorm, when each and every one of half a dozen incubating birds stood up in the hatching coops, all but one refusing to settle down on the eggs again. The very next day I went off in search of an incubator.

In selecting broodies what is really needed are the good old-fashioned breeds of hen and bantam, often with Silkie blood in them, though there are far fewer kept nowadays even in country districts. However, despite the difficulties of obtaining suitable birds some breeders, particularly the professionals who can spare the time and need every fertile egg to be hatched, still depend largely on the services of the broody, although the method is comparatively labour-intensive when compared to an incubator. If only a few clutches are to be hatched and reared then look no further, but where the number of eggs is great and time of the essence, an incubator is probably the safest bet, saving at least in time though results are seldom as good, particularly with some of the species that are more difficult to hatch.

A good, steady broody hen is worth her weight in gold when settled tightly on a clutch of expensive eggs. To test her steadiness she should first of all be given a clutch of dummy eggs for a day or two to ensure she is fully cooperative, after dusting her thoroughly with louse powder to get rid of any undesirable creepy-crawlies she may have on board, taking particular care to give a good dusting around the ventral region and under the wings where she can harbour all manner of parasites. Ridding her of the parasites will prevent her from scratching and jiggling around uncomfortably or even standing up as they start moving around, when the eggs could be damaged or allowed to cool. The trial period on dummy eggs will also accustom her to the daily routine of feeding and exercise without risk, for she obviously has to be allowed off the nest to feed, drink and exercise regularly throughout her confinement. The nest itself must be comfortable, having a deep layer of soil in the bottom of the coop to retain a little moisture moulded in a saucer-like depression to hold the eggs, and lined with a layer of soft barley straw topped up with hay. The nest itself should also receive a light dusting of louse powder. The hatching coop, of a size just large enough to accommodate the nest, must be kept completely dark inside when the hen is sitting, and the box should have a hinged drop-down door in one side to allow her on and off the nest, ideally with another hinged lid at the top so that one can check that she is sitting comfortably. Hens are best introduced to the coop at dusk, when they seem to settle down far more readily.

When taking the hen off the nest for the daily feed and exercise the whole operation should be carried out quietly and carefully with the minimum of fuss to avoid the risk of disturbing her as much as possible. When feeding she can be gently tethered by one leg to a stake on a few feet of line and supplied with food and water; alternatively let her out in a run attached to the hatching coop, possibly the better method, as after feeding she will return to the eggs of her own accord, requiring only to be shut back in once she has settled down comfortably on the eggs again with the minimum of handling to unsettle her. While off the nest the broody will also appreciate an area of dry soil where she can give herself a dustbath, the action ridding her of any parasites that may have been missed earlier.

While she is off the nest, turn the eggs and on cold days cover them with a woollen cloth or wad of duck down to help them retain their heat, and as the time for hatching approaches dampen the eggs lightly with warm water from a fine mist sprayer, which will help to create the humid conditions necessary for a good clean hatch. In the natural state a wild duck incubating eggs returns to the nest with dampened breast feathers, having fed and had a quick wash and brush up on water, and this, combined with her body heat, creates really humid conditions in the nest, at least for a short while each day before her body heat eventually dries it out. Probably for this reason duck eggs would seem to require rather more humid conditions than most other eggs, especially at hatching time, though the dampening of the eggs should not be excessive during the early stages of development because then the air space inside the shell will not be able to develop fully as the egg progresses and there will then be insufficient space for the duckling to move around in the shell when the time for hatching comes. It will then be able to chip through the shell at one point only, and being unable to turn full circle to sever the complete top off the egg will eventually die trapped inside. Insufficient moisture will bring about much the same result though for a different reason: lack of moisture dries out and hardens the inner membrane inside the shell, making it extremely difficult for the duckling to pierce with its tiny egg tooth. The duckling, small and weak from excessive dehydration, usually dies exhausted. As will become apparent with experience, getting the correct level of humidity is the key to the whole hatching process.

The broody hen needs treating with the greatest respect at all times; quiet and confident handling, a peaceful environment, and no sudden shocks, noises or movement, or she will not remain broody for long. One of my earliest and more unpleasant experiences taught me the lesson well and, although probably best forgotten, is a prime example of how not to treat the broody hen. It happened one afternoon in May.

Finishing work early, on the way home I decided to collect a batch of broodies from a local poultryman, who at the time had quite a few available in his deep-litter sheds only a mile from my home. There were several clutches of duck eggs awaiting collection at the ponds, and after sorting through what he had available the poulterer finally supplied me with eight of his best birds, which seemed to fit the bill admirably, all clucking disapprovingly, ruffling their feathers and pecking viciously when disturbed. They seemed an ideal batch. I was driving a lorry at the time, and decided to load the ricketty crate of hens inside the cab, which at the time seemed a better idea than risk upsetting them with a rather more uncomfortable ride on the back of the lorry. It was a decision I was to regret. Having travelled barely half a mile along the road I failed to notice one side of the crate coming adrift until I was almost sent over the hedge by the commotion as the flock of hens suddenly exploded all around the cab, filling the air with dust, red mite, wood shavings and something of a far more obnoxious nature. Jamming on the brakes I ground to a halt. There seemed to be hens everywhere in the restricted confines of the cab:

perched on the passenger seat, huddled on the floor, one in my lap, one clinging to the handbrake lever while yet another teetered on the driver's headrest, her claws digging painfully into the back of my neck. Not one was left in the crate. The smell was almost unbelievable. When cooped up, broody hens only leave the nest to stretch their legs and feed once a day, defaecating at the same time, at least 24 hours of waste matter accumulating inside them until they are almost ready to explode. The results can be devastating. Undoubtedly prompted by the sudden exercise, as if to some pre-arranged signal every hen seemed to choose that very moment to evacuate its load. Unable to open the windows any more than a crack for fear of their escaping, the effect was overpowering to say the least and my eyes were soon watering profusely. To make matters worse, as I continued slowly homewards I encountered a procession of horses and riders from the local stables out on a quiet country hack, and was forced to pull up on the side of the road to allow them to pass safely. The cabful of hens evidently caused great amusement as each rider on drawing level gazed in wonder until a huge grin spread across the face of each in turn. Retaining any semblance of self-respect under such circumstances was out of the question, but I did my best to look nonchalantly in the opposite direction until they had all slowly filed by. I covered the last half a mile home with all available speed. Needless to say the hens were far from broody by the time I eventually arrived home. It was a lesson I am never likely to forget.

The broody hen has now been largely replaced by the modern incubator, of which there are a bewildering number on the market, from tiny polystyrene models holding a clutch or two of eggs to huge commercial units capable of handling many thousands. Nowadays most are powered by electricity, although a few of the old large wooden paraffin incubators still survive to give sterling service.

The incubator, generally speaking, is rather less temperamental than the hen, though it lacks the efficiency of the latter, which has been doing the job since time immemorial and will never be equalled in terms of hatchability. It must be housed in a shed or room where the actual room temperature varies as little as possible, about 60°F (16°C) is ideal, for any sudden drastic changes of temperature can upset the thermostat and play havoc with the incubator's smooth running. The incubator room should be well ventilated but free from draughts, and always ensure that no direct sunlight is allowed to strike the incubator. It will never desert the eggs, though a sudden power-cut or mechanical failure can prove disastrous.

I will not delve too deeply into the mechanics of the modern incubator. Suffice it to say that it consists of a box containing a heating element. The level of the heat is regulated by a thermostat situated beside a tiny microswitch which turns the power on and off to keep the inside of the incubator at a constant temperature. To achieve this the thermostat (normally the wafer type) is designed to expand lengthways along a central rod as the heat inside the incubator rises to the required level, until it depresses the switch to turn off the power. As the machine cools down, it contracts away from the switch, allowing the power to be restored.

The incubator must always be run for at least a couple of days or so before adding any eggs, to allow it to build up heat, and so that you can check that it is running steadily at the required temperature. In the case of waterfowl this is about 102°F (39°C), a degree or so less than for most poultry and gamebirds. Besides the heating mechanism, the

Turning marked eggs in an incubator

incubator contains a water receptacle in the base which is topped up to provide the necessary humidity.

For my own purpose I chose an electric model with the capacity of hatching 100 hens' eggs, though it will hold well over that number of ornamental ducks' eggs, particularly of the small teal species. It has a clear glass top so that the thermometer and progress of the eggs can be monitored easily and without disturbance, and a slide-out egg drawer for removing the contents each morning and evening for turning. All eggs must be turned through 90 degrees at least twice daily throughout their development, preferably at 12-hourly intervals. The eggs should be marked clearly with a soft pencil or felt tip pen on both sides to ensure none are missed. Most people mark their eggs with a nought on one side and a cross on the other, but I prefer an M for morning and an E for evening, preferably in a different colour. One can then see at a glance that all have been turned correctly. Many modern incubators now have an automatic self-turning device incorporated into them, a rather useful and time-saving refinement. When setting, it is helpful to mark the eggs and keep a record of the species and setting date. This will give you a good guide to the hatching date, although the total incubation period for any species can vary by as much as a couple of days.

Whichever method is used for hatching, eggs should be inspected at regular intervals to monitor their development and to allow any that are 'clear' (infertile) or 'addled' (fertile eggs that have died) to be disposed of, therefore avoiding them taking up precious space in the incubator, or turning bad and emitting putrid gases as they begin to decompose. 'Candling' the eggs, providing one knows what to look for, can be carried out quite easily and accurately with a little experience. The first inspection is best carried out at about seven days of incubation when the eggs can be tested for fertility. The first signs of the duckling's development can easily be seen by holding the egg over a strong light in a darkened room. Proper candling torches can be purchased, but are rather expensive. A cheaper method is to place a small cardboard or wooden box over a high-wattage light bulb (Fig. 9), blocking out all the light except for a small egg-shaped hole cut in the top, slightly smaller than the egg, on which each egg can be placed directly over the bulb. The light showing through the egg reveals the contents. Eggs that show no sign of life or development are known as clears, and should be removed to give more space in the incubator. In the case of a fertile egg, the contents will rise to the uppermost part of the egg and a tiny red, spider-like embryo will be seen, which in a few days will show the first signs of movement, the beating heart and regular kicking

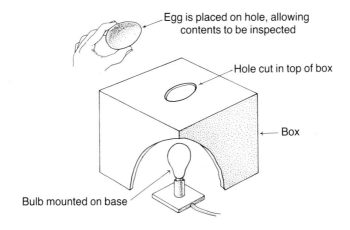

Egg is placed on hole, allowing contents to be inspected

Hole cut in top of box

Box

Bulb mounted on base

Fig. 9   Egg candling

movements easily visible. Candling should be repeated at about 14 days, for some of the embryos which looked strong and healthy on the first inspection may yet still die, and as duck eggs decompose quite quickly they are better safely removed out of harm's way.

As the embryo develops, the egg contents will gradually darken to make inspection difficult, but given the correct level of humidity the air space at the larger end of the shell will be seen to slowly increase until it ideally takes up almost one third of the shell by hatching time, when movement again becomes visible, the membrane separating the duckling from the air space begins to move as the hatchling starts to cut its way through. A regular inspection of the air space is probably the best indication of whether or not one is maintaining the correct level of humidity, as it varies with fluctuating weather conditions. Eggs are best checked every

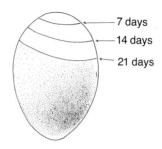

Fig. 10   Development of the air space in an egg

seven days to see that the air space is developing correctly. Figure 10 gives a rough indication of what to look for.

At seven days the air space is little more than a visible cap at the larger end of the egg; but a week later it should have almost doubled in size. At three weeks it should have increased to take up at least one quarter of the egg. Any marked variation from this can be regulated by providing higher or lower humidity by filling or emptying the water tray accordingly.

Once the ducklings begin to hatch, the eggs will produce a hollow rattling sound while being turned. They have then almost reached the 'chipping' stage, and within hours a gentle but insistent tapping will be heard as the duckling begins its struggle from the shell, first tapping a small hole towards the larger end of the egg, and then slowly revolving around the top and breaking through the shell using the tiny egg tooth situated on the top of its bill, cutting a trapdoor to the outside world through which, if all goes well, it will finally push itself free of the shell (Fig. 11).

As soon as a batch of eggs have begun to chip in the incubator they should be transferred to the hatcher. I use a

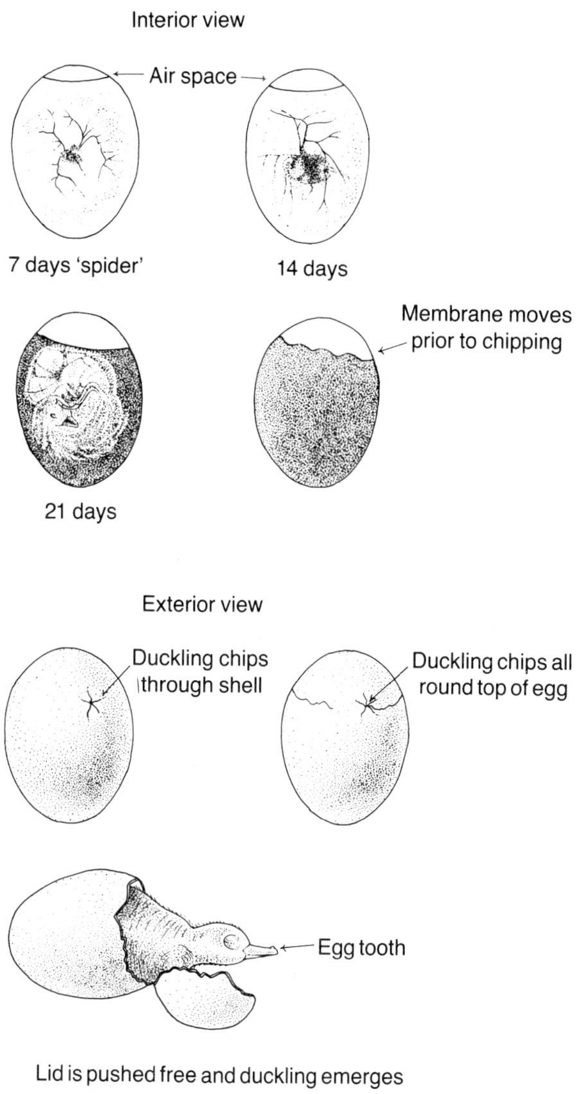

Interior view

Air space

7 days 'spider'

14 days

Membrane moves prior to chipping

21 days

Exterior view

Duckling chips through shell

Duckling chips all round top of egg

Egg tooth

Lid is pushed free and duckling emerges

Fig. 11   Development of an egg

cheap polystyrene incubator for this purpose. There is a sound reason for removing them from the incubator. In the process of hatching, eggs generate a lot of heat, and if left among others that are in various stages of development, many can perish as the temperature inside the incubator is often pushed up by several degrees. If there is only a single batch of the same species all well and good, but if the incubator contains mixed clutches of all types and ages, as mine does, a separate hatcher is essential. The eggs should be quickly transferred, placed carefully in the hatcher with the 'chips' uppermost, and dampened thoroughly to soften the shells and create a really humid atmosphere. A small, plastic plant spray is ideal for this, producing a fine mist that creates an ideal environment for hatching. Never be tempted to interfere while eggs are in the process of hatching. If you open the hatcher, much of the humidity will be lost. In addition, sudden changes of temperature are to be avoided. Leave them to their own devices for at least 24 hours after chipping, by which time most should have hatched out safely, but if the ducklings have not started cutting around the shell by then there is a chance that all is not well, and assistance may be required to help them break out of the shells. It is always difficult to know just when to intervene. If too early, the yolk sac joined to the duckling will not have been fully absorbed. If too late the duckling, now growing rapidly, will have started to outgrow its shell, and even if rescued alive is likely to be deformed or crippled from being cramped inside the shell. To test the readiness of any reluctant eggs, carefully remove a small portion of the shell from the point at which the egg has been pierced by the duckling to reveal the state of the tough membrane just below the surface of the shell. If the membrane is clean, white and unbroken it is probably too early to interfere, for

Newly hatched Scaup ducklings

the membrane will bleed if punctured too soon and can result in a dead duckling. If, however, the membrane appears discoloured and somewhat burnt-looking and the duckling shows little sign of progress, it is certainly time to help, but stop immediately any sign of bleeding occurs. At this stage it is usually a matter of salvaging what you can rather than hoping to save the entire clutch. Any youngsters that are crippled in any way are best put down immediately, for they will certainly not improve however long one keeps them.

When assisting an egg to hatch, carefully remove a thin line of shell along the same line the duckling would have taken, and gently peel away the membrane to ensure the lid will come off easily, though stop immediately any bleeding results. Once the lid is removed replace the egg in the hatcher with the bird still inside. This will give it a little space to stretch in, and if all goes well the duckling will rest for a while before kicking free of the shell, allowing time for the last traces of the yolk to be absorbed. On no account forcibly remove a bird completely from the shell, for the chances are that an unabsorbed yolk will either rupture or stick to the floor of the hatcher as the bird dries out.

Once safely hatched, allow the ducklings a few hours to rest from their labours and dry out completely in the hatcher before removing them to the next stage, the rearing unit.

9

Rearing

U<small>NLESS ENLISTING</small> the services of a broody hen, directly a batch of ducklings have fully dried out in the hatcher it is time to transfer them to a brooder for rearing. A duckling's requirements are few – food, water and constant warmth are all that is needed to rear a tiny hatchling into a vigorous and healthy, fully-fledged bird. Hatched out under a hen, only food and water are required, for the foster-mother will brood and warm the ducklings. Provided that they are housed in a secure pen of small mesh to prevent them straying and to protect them from the attention of stray cats, rodents and other dangers, she should experience little difficulty in rearing them.

Hatched in an incubator, the ducklings will need to be provided with some form of heating right from the start, and although I began with a system of cheap paraffin brooders which carried me through the initial seasons, this method has now become rather outdated. I have switched to electricity, a far easier, cleaner, maintenance-free and very successful alternative. The paraffin brooders served their purpose at the time, though constant vigilance was required to keep the system running smoothly. Small heating stoves

141

were lit and placed beneath the wire-mesh floor of the brooding unit, the heat from the flame rising through the floor to keep the brooding-box warm. It was extremely difficult to keep anything like a constant temperature, as the wicks needed constant attention to keep them burning evenly and prevent them smoking, which would eventually soot up the inside of the heater and extinguish the flame. The brooder-house permanently reeked of paraffin and there were frequent problems with watery eyes – not only among the ducklings.

The main drawback with using electricity is the chance of a power-cut or a bulb exploding, both of which can prove disastrous. The latter is sometimes caused by the bulb being splashed with cold water as the ducklings drink and preen. For this reason the heating bulb must be placed well out of range of the water dish. Unexpected power-cuts one can do nothing about, but provided heat is restored within a few hours most ducklings show amazing powers of recuperation. All in all, the merits of electric heating far outweigh the occasional disasters.

With a simple overhead electric heating bulb and reflector, a brooder can be set up safely almost anywhere: in the penned off corner of a shed, inside a large box, or in a circle of heavy-gauge small-mesh wire netting, which can be overlapped upon itself at first, then gradually widened out to provide more space as the ducklings grow. The heating lamp can be suspended from the ceiling, or from a convenient rafter on a length of chain. This allows the level of heat to be regulated easily as the ducklings progress by adjusting the height of the lamp a few links at a time, raising the bulb to give less heat and allowing the ducklings to harden off gradually, until at about three weeks of age they have become acclimatized to normal temperatures. The advantage of using a light bulb is that ducklings are encouraged and able to feed during the entire 24 hours of the day, and their rate of progress improves accordingly, hastening development quite noticeably compared to the old system. At three weeks of age most ducklings will have begun to sprout the first stubby signs of feathers on their flanks and back, and by this stage little or no heating will be required, except on very cold days or if the ducklings have become wet and bedraggled from bathing.

When ducklings are first transferred from the hatcher to the brooder situate the heating bulb quite close to the floor, giving a maximum temperature of about 90°F (34°C) in the centre of the circle of light, a temperature which is rather too high for most young fowl. In this way the youngsters will be encouraged to seek a spot where the temperature is most comfortable. They will form a ring around the light at the desired heat, which prevents them from all crowding together in a huddle if the heat is insufficient. While they need constant heat for their first few days of life, by a week old the level can be reduced. On warm days it can be switched off altogether during the hottest part of the day, although always remember to switch it on again in the evening, for even in full summer the temperature can drop drastically during the hours of darkness.

The brooder floor can be covered with peat or wood shavings or even sand, though the latter can ball up on their feet uncomfortably, especially when the ducklings have grown sufficiently to produce large amounts of waste matter. Straw should on no account be used because it quickly becomes dirty, wet and mouldy, as I once found out the hard way, creating ideal conditions for all sorts of problems, among the worst of which is aspergillosis, a respiratory problem caused by a form of mould whose spores are present in mouldy bedding. Infected ducklings show a gradual loss of condition over a few days and eventually their breathing becomes laboured. By this time the infection is well advanced and death occurs shortly afterwards. There is no treatment for the condition, which almost always has fatal consequences.

Wood shavings or peat will absorb most of the droppings at first. But as the ducklings grow, regular topping up will become necessary to keep the floor of the pen in a reasonably hygienic state. For this reason, a couple of seasons ago I decided to completely revise my rearing system of makeshift brooders, for as the numbers and types of fowl in the collection has increased, the number of youngsters reared annually has risen correspondingly, and more time than could be spared was needed to care for them properly. There is little pleasure in keeping birds under foul conditions, so I set out to install a system which was both labour-saving and efficient, and above all hygienic. In the end I settled for a method using raised wire pens–an idea copied from a professional breeder friend – which eliminates the need for constant cleaning out and results in much healthier and better plumaged juveniles (Fig. 12). Raised 3 ft (90 cm) from the floor on wooden legs, each individual brooding unit measures 3 ft (90 cm) square by 18 in (45 cm) high. The wooden framework of 2 × 2 in (5 × 5 cm) timber is covered with heavy-gauge ½ in (1.25 cm) square welded wire mesh, including the floor, which allows all waste matter, droppings, spilt food and water, to drop straight through the wire to the concrete floor where it can be cleared up easily with a broom and hosed down from time to time.

The row of ten units, at times housing well over 100 ducklings, can all be cleaned up in one fell swoop, and the

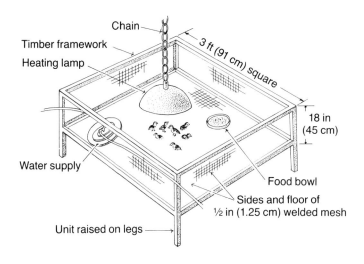

Fig. 12   Raised wire brooding unit

whole task of mucking out now takes only a matter of minutes. In this way the brooding area is kept clean at all times, even when the ducklings grow large and produce a correspondingly large quantity of waste matter. There is also no danger of the ducklings eating spilled and contaminated food which has turned mouldy underfoot.

In each pen I have installed an automatic drinker, another great saver of time and effort. The simple shallow plastic bowl is fed by a small tube worked by a valve which operates to control the water level. The line of drinkers is supplied by gravity from an overhead tank kept topped up by a ballcock valve. The complete system was very cheap and simple to install and now, even in very warm weather, when the need for water increases dramatically, there is a clean and constant supply.

145

Ducklings in a raised wire brooding pen

With wire floors there is also the great advantage of never having to top up or renew bedding material when it becomes fouled, and my main rearing duties are now confined to topping up the food bowls and checking that the water supply is working, allowing more time to be spent actually enjoying the fowl instead of rushing around at least three times a day attending to their continual needs.

Hand-reared geese are the exception. Not ideally suited to the system, the few I hatch in an incubator are best installed in an outside pen. Being natural grazers they are far better reared on grass in a small moveable pen with a hut attached, again heated with a bulb and reflector, where they can graze their fill and be moved to fresh grass each time the area becomes soiled. They can be started off in a brooder for the first few days, and fed on chick crumbs to which some finely

Goslings in a moveable pen on grass

chopped grass has been added. But if they are kept as such for too long they can develop leg problems, and with no grass to keep them occupied in grazing, they will often resort to feather-picking one another.

My duckling brooders are built along one side of a 30 ft (9 m) brooder shed, and in the corner of each individual brooder there is a small pop-hole which allows access to an outside pen of the same construction, 10 ft (3 m) long and also raised on legs, the pop-hole door being raised or closed like a miniature portcullis by a length of nylon cord running through a staple above the pop-hole to allow it to be operated from the far end of the outside pens. It is therefore a simple matter to shut the birds in on a cold night or if a sudden storm is expected. The system allows young fowl to spend much of their time outside in the fresh air if the day is fair, or to retreat inside the hut to warm themselves under the lamp if the weather turns cold or they get caught in a sudden downpour. They will enjoy a light shower, and a little rain works wonders for their plumage, encouraging them to preen their feathers and acquire a healthy sheen of oil that will soon withstand and repel even the worst of weather. Contrary to popular belief, ducks are not automatically waterproof and if reared without access to bathwater, will soon become waterlogged, and in extreme cases can even drown, on their initial introduction to a pond. If reared on wire, the birds can be given water in a shallow tray in the outside pen, where they can swim and bathe at an early age, for the pens will always remain dry underfoot, and the ducklings will warm and dry themselves beneath the lamp afterwards, becoming completely waterproof after just a few days of this treatment.

While they will enjoy being able to get out in the fresh air as soon as possible, during very hot, sunny weather one must ensure that the pen has a shaded area to which the ducklings can retreat from the full power of the sun, for until fully fledged they can be susceptible to sunstroke on very hot days. Part of the pen can be covered with hessian sacking or a few leafy boughs to provide shade during the hottest part of the day.

Having rigged up a suitable brooder, one has to decide how to prevent the youngsters' future escape from captivity. Obviously, birds cannot be left full-winged or they will almost certainly stray, unless conditions are really favourable. Pinioning is the obvious answer and, carried out at an early age, is an easy, straightforward and painless operation; it is also permanent and effective. It is simply a matter of removing, from one wing only, the tiny end joint that later holds the primary feathers of the fully-fledged bird.

Raised wire pens built outdoors along one side of a
brooding shed

This has the effect of unbalancing the bird should it attempt
to take to the air, and renders it incapable of flight. The only
other alternative is wing-clipping, which involves clipping
the actual primary feathers on a regular basis to achieve the
same result. This has the obvious disadvantage of needing to
be carried out after each successive moult, when every bird
in the collection has somehow to be caught up and handled.
Wing-clipping can also be rather unsightly, whereas one
hardly notices the absence of primary feathers on one side of
a correctly pinioned bird.

Whereas some breeders leave the ducklings until they are
at least a few days old, I prefer to pinion at the earliest
possible opportunity during the move from hatcher to
brooder. This means that the task is over and done with
permanently, and once the birds have settled in their new

home and are feeding well there is no further need to risk upsetting them by catching up and handling, both of which causes unnecessary stress. At this early stage, a scalpel and block, or sharp pair of scissors will do the job painlessly and quickly. Holding the duckling gently but firmly in the palm of the left hand, with its head facing towards you, grasp the bird's outstretched left wing between thumb and forefinger as the hand envelops the body. Insert the scissors carefully against the alula – the tiny limb that projects halfway along the wing – and keeping the blade of the scissors as near to it as possible, remove the end joint neatly at its axis (Fig. 13). It is most important to leave the alula on the wing, as it covers and eventually protects the point of amputation. Carried out correctly, pinioning causes little distress, and the ducklings suffer no ill effects from the experience.

A word of warning. The ducklings of tree hole nesting species are remarkably adept at climbing out of a brooder as they possess an ability to scale unbelievable heights by clinging to wood or wire with their tiny claws. Normally hatched in a deep hole inside a tree trunk, as soon as they have dried out after hatching, their first instinct is to climb upwards to escape the nest. This behaviour persists for the first couple of days, when the brooder should if possible be covered. Mandarins and carolinas are particularly troublesome and often spend the first day or so doing their utmost to escape, until the instinct finally wanes.

When putting them into the brooder for the first time, dip the beak of each duckling into the water bowl. This shows them where the water is. All water bowls must be shallow, and as a precaution put a few small stones in the bottom so that any duckling unfortunate enough to fall in will be able to get a grip on them with its feet and clamber out to safety.

Newly hatched ducklings will rarely feed much during the

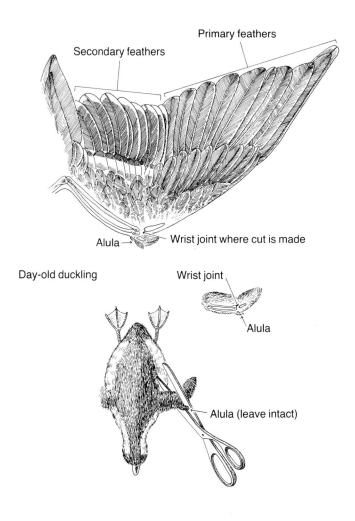

Fig. 13   Pinioning a day-old duckling

first 24 hours, as they are still digesting the remains of the yolk inside them, but every effort should be made to get them taking food as soon as possible to get them off to a good start in life. Once they are feeding well, half the battle is over. In the wild all young waterfowl quickly learn to feed themselves by copying the actions of their parents, but those hatched in an incubator can sometimes cause problems, especially the ornamentals, many of which are fickle feeders and often need a little encouragement to get them going. There are several tricks of the trade to start them feeding. As soon as a batch of ducklings are settled happily in the brooder, sprinkle a handful of chick crumbs on newspaper spread on the brooder floor, for as they shuffle around the ducklings will send little pieces rattling across the newspaper and the noise and movement will often prompt them to peck at it. A handful of crumbs sprinkled over the ducklings' backs will often encourage them to preen it off one another, and a few crumbs floating in the water dish may also help, together with more beak-dipping. In extreme cases I have even suspended a can of water over the water bowl, the base of which is punctured with a hole that is just large enough to allow a tiny drip of water to fall every few seconds into the dish below. The splashing and movement can work wonders, causing even the most stubborn to investigate. Anything that moves or wriggles is far more likely to attract attention, and as all wild ducklings are insectivorous for their first few days of life, a few meal-worms in the food bowl may prompt them to begin feeding. If all else fails, put a slightly older duckling that is feeding well amongst them to teach them to feed, for the youngsters will soon learn to copy its actions as they would their natural parent. Once they have learned, the 'teacher' can then be returned to its original brood.

Most of the common species of duck can be reared at first almost exclusively on a basic diet of chick crumbs, though all ducklings love a little finely-chopped hard-boiled egg, and mealworms and chopped greenstuff can be given as an occasional treat to keep up their appetites. At the earliest opportunity, gradually wean them to small growers' pellets, for if crumbs alone are fed for too long it can lead to problems in their development.

When I first began rearing ornamentals I noticed several unexplained cases of wing droop during the feathering stage, where the wing seemed to grow too heavy and in some cases was soon dragging on the floor. Apparently the high-protein content of chick crumbs causes this, as too high a level is capable of damaging the kidneys and depositing urates in the joints, which can produce wing droop and also leg weaknesses, another all-too-common ailment of young waterfowl. Mild cases of wing droop can usually be cured by taping the afflicted bird's wing in its natural position to support the primaries, leaving the joint bound up for a maximum of a week. By this time the treatment should have worked, but in any case the wing should not be left taped up for longer or lasting damage could result.

Feather-picking among growing fowl can be a problem, although some species seem more prone to it than others. It is a most annoying condition and can be difficult to prevent once started; in extreme cases it can lead to the disembowelment and death of a victim. Boredom appears to be the main cause, and among the ducks, red-crested pochard and wigeon seem the worst offenders, particularly if large numbers are confined in too small a space. Many of the geese are also liable to feather-pick unless kept occupied on a sward of fresh grass. A plentiful supply of greenstuff helps alleviate the problem, but once started it can prove the very devil to eliminate except by isolation of the offenders.

Having reared a batch of young waterfowl to maturity, a period of some eight weeks or so, they will need to be sexed before being offered for sale. Even in eclipse plumage, most of the adult fowl of the northern hemisphere are readily sexed, the more colourful plumage of the drakes allowing instant visual sexing from their more soberly dressed partners. In the case of many southern-hemisphere species, the sexes are almost identical and although there are usually small differences between them, the only safe way of sexing accurately is by internal inspection of the vent. The same applies to youngsters of both types before they acquire their adult dress.

Vent sexing will enable one to dispose of youngsters long before they assume their full colours in the autumn, for although in some cases there are ways of separating duck from drake, it needs a practised eye to differentiate with any accuracy. The redder bill of a young mandarin denotes a drake, as does the facial stripe of a young carolina, or in some cases the more highly coloured iris of a young drake's eyes.

To make absolutely certain, vent sexing is the only answer. Catch a bird and turn it on its back with the head pointing towards you, holding the body gently between your knees to avoid it struggling to escape from what is a rather undignified posture as you smooth back the short feathers to expose the vent. If the bird quacks as you pick it up it is almost certainly a female, but to make absolutely sure a closer inspection is needed. Using the thumb and forefinger of each hand, gently and carefully peel back the folds of the vent exerting firm but gentle pressure towards the tail. This will eventually expose the sex organs. Almost certainly this is not all that will be revealed, for invariably the duck will choose this moment literally to vent its annoyance at being so treated. Brush the resulting mess aside and patiently continue to unfold the vent until, in the case of a drake, a tiny penis that can be likened to a sprouting pea will appear, which is obviously not present in the female. It takes a little practice, but with experience the job can be carried out in seconds. After sexing, the bird can be marked by clipping 1–2 in (2.5–5m) from the end of a couple of primary feathers to denote one sex, the opposite gender being left unmarked. This will enable fowl to be recognized at a glance when you eventually catch them up for sale, which saves much in time and unnecessary handling.

Hatching Period of Waterfowl Eggs

These periods are approximate, as length of incubation can vary by a day or so either way according to weather conditions and individual birds.

Swans	*Days*
Mute	35–42
Whooper	35–42
Bewick's	35–38
Black	35–37
Black-necked	35–36
Coscoroba	35
Trumpeter	35–40
Whistling	35

Geese	
Greylag	28
Pink-footed	28
Bean	28
European white-fronted	28
Lesser white-fronted	27–28
Greater snow	26
Lesser snow	23
Ross's	23
Bar-headed	28
Emperor	26

Swan	28
Barnacle	25
Brent	25
Atlantic Canada	28
Red-breasted	25
Hawaiian (Ne-ne)	30
Egyptian	28–30
Ashy-headed	30
Ruddy-headed	30
Andean	30
Abyssinian blue-winged	30
Cape Barren	35

Ducks	
Mallard	28
Gadwall	25
European wigeon	25
American wigeon	25
Chiloe wigeon	25
Northern pintail	23

Bahama pintail 25
Northern shoveler 24
Argentine red
    shoveler 25
Mandarin 30
Carolina 30
Australian wood
    duck 28
Shelduck 28
Ruddy shelduck 30
Australian shelduck 32
European green-
    winged teal 23
Blue-winged teal 23
Marbled teal 26
Ringed teal 23
Cape teal 23
Versicolor teal 25
Hottentot teal 20–22
Chilean teal 26
Chestnut-breasted
    teal 27
Falcated teal 25
Baikal teal 24
Cinnamon teal 25

Laysan teal 28
White-faced treeduck . 28
Fulvous treeduck 25–26
Tufted duck 25
Ring-necked duck 28
European pochard 26
American redhead 26
Canvasback 26
Red-crested pochard ... 26
Rosybill 24
Ferruginous 26
European greater
    scaup 26
New Zealand scaup 24
Lesser scaup 22
European goldeneye ... 28
Barrow's goldeneye 30
European eider 28
Hooded merganser 30–32
Goosander 30
Smew 30
North American
    ruddy 23
Argentine ruddy 23

Index